UNDER THE NURSE'S CAP

NURSING SCHOOL YEARS

RN - BSN

By:
Henry Intili, RN, NP

Copyright 2008

OTHER BOOKS BY HENRY INTILI

REGULAR BOOKS:
Backpack and Canoe: Explore the Beautiful Places
Backpack and Canoe 2
Young and Single in NYC 1965
Hippie Years on the Farm 1978
The Trigamist
Anello and the Garibaldi Reunion
Anello and the Soldiers Returned From World War 2
The IMP Affair
The Weenie Clinic: Two Years in a Mens Sexual Dysfunction Clinic
Recipes From Gloria's Restaurant in NYC
More Recipes and Stories From Gloria's Restaurant
The Adventures of Tony and Woof
More Adventures of Tony and Woof

eBOOKS:
Exploring Alaska by Canoe and a Rent-a-Wreck Plymouth Van.

A bicycle trip in Holland through the tulips – Leiden to Haarlem.

A hike in Iceland, the Thorsborn trail

A walk in Italy, Tuscany from Florence to Siena

A hike in the Canadian Rockies, the White Goat Wilderness, Banff National Park

A walk in Ireland on the Dingle Way

A canoe trip on the North Fork of the Koyukuk River, Gates of the Arctic National Park, Alaska.

A canoe trip on the Missouri River, 100 miles in Montana.

Canoe the Noatak River, Gates of the Arctic National Park, Alaska

A hike in Needles, Canyonlands National Monument, Utah.

A hike on the Brazeau River Trail, Jasper National Park, the Canadian Rockies.

Walks with Ding-Dong and other Poems.

Details at www.henryintili.com

DEDICATION

In 1982 in Albany, New York I went to a palm reader named Mother Antares. Scorpio is my sign and Antares is the red star that is the heart of the scorpion. Those days I was a laboratory equipment salesman. She told me that I would have a second career in the medical field and marry a woman with children named Steven and Stephanie. I thought I had wasted twenty dollars on a foolish whim.

Seven years later I met Barbara and we married exactly a year after our first date. She has children named Steven and Stephanie. And two years later Barbara agreed to let me put aside my sales career and go back to school at the age of 45 to become a registered nurse.

My career as a registered nurse and later as a family nurse practitioner would not have been possible without her love, trust and support.

Henry Intili, RN, FNP
PO Box 125
Bainbridge, Georgia 39817
www.henryintili.com

BEGINNINGS

1967 and missed medical school

When I returned to Dickinson College in Carlisle,
Pennsylvania with JoAnn in the fall of 1966, I enrolled as a Pre-
Med major for my senior year. In that year of the escalating war in
Viet Nam, the option for healthy boys graduating from college was
to continue in some kind of exempted graduate work or be drafted
or enlist in the army. Not much of a choice in my book. I hadn't yet
begun to march and demonstrate against the war - that would come
later. But I sure knew that I didn't want to participate in an armed
struggle in Southeast Asia.

In the spring of 1967 The Medical College of the University
of Pittsburgh asked me to visit their campus for a day of interviews
and discussions. JoAnn spent the long drive across the state
knitting a sweater. She was always knitting sweaters. Everyone
loved her sweaters.

I did not keep a diary in 1967. The memories of that trip to
Pittsburgh are distorted and selective so that only intense
fragments remain. Sweaty palms and jellied legs as I watched a
med student carving on his cadaver and holding a detached foot up
in the air. Stomach queasiness as I watched a live operation. JoAnn

was fascinated, unaffected.

It was fear that controlled my subsequent actions. They accepted me for admission in the fall of 1967. I turned them down, telling JoAnn and family that I needed some time away from school. Did anyone believe the line? Did I? Ridiculous.

Two weeks later the Selective Service Board of Essex County, New Jersey called me for my induction physical. I was in the bloom of perfect physical health, no way could I fail. Here was my choice: two years as a draftee in the army or three years as a volunteer enlistee in officers' candidate school. I called OCS and enlisted. I still had to take the physical.

They rejected me. Classified me 1-Y, exempt until time of declared war. It was a day long physical of multifarinous pokings and probings with a hundred other boys, many my high school classmates. At the end of the day we were assembled in a large room. Facing us, seated at a long folding table were three bored physicians dressed in rumpled lab coats. A sergeant, hair cut to within a half inch of his skull and meticulously dressed in starched browns with rows of ribbons over his breast pocket, stood to the side calling out names. He came to mine.

"Intili!"

"Here," I answered walking to the front for my ticket to the jungles.

"We find you unfit for military service."

I stood motionless, silent.

"I repeat: We find you unfit for military service."

I stood motionless, uncomprehending.

"Son, get the fuck out of here."

I left. They never told me why I failed. I never asked. I wasn't going to Viet Nam. And I didn't have the slightest idea what I was going to do.

The failure to attend medical school was only the start of a string of destructive decisions that included the loss of the love of that wonderful girl. In those two years between 1967 and 1969 I squeezed the toothpaste tube of life opportunities until what remained was a twisted, empty hulk.

1980 and the origins of nursing.

In 1979 it became increasingly clear that the farm
life/experiment high on a hillside in rural Western New York was
failing. I tried a number of remedies to escape the sticky strands of
my own spider web. At one point I put together a grand scheme
involving a move to the nearest town, Hornell, where I could open
a small woodworking shop and go back to school at Alfred. I
learned that Alfred State College had a nursing program and went
for a visit.

Although I was keeping a journal at the time, for some reason
I do not mention this trip to the nursing school. Did I even tell
Yvonne (second wife) I had gone? Probably not, considering the
general state of communication between us that year. In any case,
my memories of the day are plastic. I spoke with a white starched
female professor who was polite but discouraging. "A male, and
excuse me for being blunt, an older male at that, will have a
difficult time in the program and a difficult time being accepted by
other nurses. However, I cannot legally stop you if you want to
apply."

Application became pointless when I could not figure out
how to merge the declining finances of the household with the
demands of a return to school. Or maybe I was too proud to
consider nursing after having blown the opportunity for medical

school ten years earlier. I'll never know the truth of my feelings in that cacophony of failure, rejection and pain that limns 1980.

March, 1991.

In the spring of 1991 when the job as National Sales Manager for Medical Products at Neotronics in Gainesville fell apart even though I had made the first substantial sales of their products in the United States, I decided that I could not continue to work at jobs where I could be easily fired. The loss at Neotronics drove me to a near panic. I could not permit my life to slip into a repeat of the financial chaos that engulfed me in 1987 in Oak Ridge, Tennessee when Unlimited Brokers of Equipment went belly up.

The final nail in the 1991 job coffin came when I learned that the president at Neotronics would not give me a recommendation. I had to bury any thoughts about professional level work. That's why Barbara and I started the woodworking business doing small jobs for Marriott. I also began work as a salesman with Morgan Buildings and Spas that spring. I was willing to do anything to bring in a dollar and not repeat New York State in 1980.

Barb and I sat down to chat on a swing in our Atlanta backyard that was bathed in lambent evening light.

"When I was working for Neotronics, I really enjoyed working with the registered nurses. I was wondering what you thought about my going back to school."

"To become a registered nurse?"

"Yes."

"As long as we can find a way to bring in enough money to pay your share of the rent and groceries, I say: Go for it. My work at Marriott will cover us for the rest."

Several weeks later I set up an interview at the Nursing Department at Kennesaw State College. On the fourth floor of the library, in a tiny cubicle office, I met Professor B of the Nursing Department.

"How can I help you," Professor B asked in a pleasant, friendly voice. He was about forty years old and supremely comfortable in an advisory role.

"I don't know if you see male applicants or especially older male applicants."

"More than you would guess. We welcome men. This year we're 15% men in the program."

"I want to explore attending the school of nursing. I've been in the medical sales field for many years and I think it's time for me to jump the fence, so to speak." Then I went on, in a convoluted way, to talk about using a nursing degree to help me in business. I was willing to say anything to preserve my dignity, keep me above what I envisioned as a classroom of little girlies with their white caps.

In return Professor B gave me the standard introduction speech with an explanation of the two RN programs (two year AD and four year BSN). I didn't listen; my mind was flooded with images of business failure, deceit filled chaos, sickening embarrassment. On the way out a large, friendly secretary named Maggie provided me with explanatory flyers and admission forms. I drove home obsessing in dread.

Kennesaw State Uuiversity and Mickey Mouse Core Courses.

I filled out the various forms and submitted notarized and sanitized transcripts from my college work thirty years earlier. The good news was that I was able to supply these resurrected, ancient papers. The bad news was due to differences in curricula between Dickinson College (1967) and Kennesaw State (1991), I would have to take a number of undergraduate courses outside the nursing sequence and before admittance into the nursing program in order to meet the "core" requirements for graduation. Physical Education, English World Literature, and Human Development struck me as particularly silly which seemed in line with the whole college process.

In order to register for classes I had to provide proof that I had been vaccinated against measles. I called the Registrar's Office and spoke to a lady who sounded about 18 years old.

"I'm 45 years old," I said. "I had the disease as a child. We all did in our age group."

"Can you get a letter from your childhood doctor?" she asked.

"That was 35 years ago. Who keeps records for 35 years? And he's probably been dead for twenty years or retired."

"We could accept a letter from your mother."

"My mother! She's been dead for 15 years." I wasn't thinking or I would have accepted her offer and written a letter under my mother's name. Perhaps even give the return address as Queen of Heaven Cemetery, Fort Lauderdale.

"Well, the rules say you have to have proof of immunity."

Later that evening Barbara listened to my tale of woe and booked me into an appointment at the Fulton Health Department. That's how Barb solves problems: she takes them by the horns and throws them. Instead, because Kennesaw State (KSC) is in Cobb County, I went to Marietta and visited that health department.

A starched RN came out to talk to me. "You need an immunization for measles?"

"Yes. For college admission."

"You're over 35. You've been exposed as a child. We don't give immunizations to people over 35. There's no point to it."

"I know that, but the college insists."

"I'm the County Health Nurse and I insist that you don't. Do you have a phone number I can call to set these people straight?" She was about 40 years old and filled out from age. I was impressed with her attitude and skeptical about her chances.

I listened as she made the call in her office. Her voice started low and professional and slowly rose in timbre and volume. The phone slammed down and out she stormed. "Idiots!" She shook the moths out of her brain and looked at me. "Which arm do you want the shot?"

In the Human Development course that I read the chapters every night and answered all the questions at the end whether or not they were assigned. The instructor told the class that he could pick out the nursing students: "You're always focused."

In Physical Education the young kids ran circles around me. But they didn't bother to find out what the semi-retired professor wanted from his students. He was more interested in class participation than in demonstrated physical prowess. I got the A, they got the lower grades. Age and deceit always beat youth and talent.

World Literature was a pleasant surprise. The professor was from England, educated at Oxford and in possession of depth in his material that I rarely encounter. He challenged the class every night. He was often frustrated in his efforts to wake the sleeping dead.

"Miss Jones," he asked a particularly vapid girl with brilliant blond hair and a creamy complexion, "How do you rectify the picture of a vengeful God in Leviticus with Christ's preaching of love and peace in Luke?"

"I don't," she replied without sarcasm.

"Pardon me?"

"I don't think about it at all. That way it doesn't bother me."

"Right," he said turning around and walking toward the black board. Did I see him shaking his head? He left Kennesaw a year later to teach in North Carolina.

Pre-Nursing Courses.

That first year I took two courses in the pre-nursing sequence: Pathophysiology and Nursing 208 (Introduction to Nursing). Patho was a joint teaching effort between a Biology Professor and a Nursing Instructor. A solid structure with cracker-jack teachers. I soared in the material, leading the class in most of the tests. During one test review I questioned the professor about a particularly arcane point. "I have a problem with the answer to question six. Wouldn't the acidity be reduced when...."

"Henry."

"Yes, Professor?"

"What do you care? This is probably the only question you missed this semester. Even if you're right, it won't matter. Give it a rest."

In January 1992, I was not sure if I could continue the pre-nursing sequence. Barb and I were nearly broke. My job at Morgan Building had only lasted until September when the sales dropped off. Although the woodworking volume in our small shop had picked up, the gross profit was low and my elbows were in constant pain. Barb and I were on slippery financial ground from this and a new mess I had embroiled us in.

UBE and the Gas Centrifuge Plant in Portsmouth, Ohio sell-out from 1989 had reared its head again from the legal morass that had closed us down. The owner wanted me back to selling the assets from the plant as I had done three years earlier. Dianne dangled promises of big money and bonuses in front of me in November. Initially I would have to lay out some of my own time and resources until the money from the sales rolled in; then all would be repaid and more. Except that due to government snafus, nothing could be sold or shipped until some "details" were worked out. I agreed to this crazy scheme.

By January Barb and I had sunk several thousands of our own dollars into the project and were living in a panic about losing everything. School and nursing were being squeezed out by the cash crunch. I couldn't see how I could do school that year.

I called Professor B and explained that due to financial problems, I wouldn't be able to continue in the program this year.

"That's disappointing," he said. "We were looking forward to having you around."

George, my old boss at Morgan, asked me to come over and build two decks for him. January 8th was a sunny, warm, Georgia winter day. Perfect for outside construction. I stopped for lunch, a sandwich on the half-finished deck. My elbows ached from sawing and drilling.

17

I looked out over the Morgan parking lot, my mind rolling in panic and terror over finances and future doom. And suddenly, in an epiphany, I knew that I had to get back in the nursing program. I could not repeat 1967. I could not sit out my positive future for present fears. I had to do whatever was needed, financially or otherwise, to stay in a positive direction. Treading water, delaying progress - these were repeat warrants on my future.

The next day Barb and I drove to Kennesaw to meet with Professor B at 1 pm. We arrived early to eat lunch in the cafeteria, and in the hot food line we met him. We three sat at an institutional table with plastic chairs. I was ready for a long sales pitch. He didn't even let me start.

"No problem with the add slip for Nursing 208 from me. After lunch we'll go to my office so I can write one out for you. I'm the RN Program Chair. But classes have already started, so you'll need a signature from the registrar's office."

I should have known. The Registrar's Office again.

"I need a signature on this add slip," I told the lady at the front desk in the Registrar's Office who looked through me to some spot on the back wall. Hello, am I invisible?

"The Director of Registration needs to sign that. You'll have to wait." She pointed to a line of ten people.

"I already have the approval of the Department Chair."

"The rules state you must have the Registrar's signature after classes begin."

Barbara and I sat. Half an hour passed. I was pissed. Inappropriately angry, I walked back to the bored lady. "Miss, I've been here 30 minutes. This line is hardly moving. What's the problem here?"

"There's nothing I can do. You have to wait for the Registrar."

An hour later he called me into his office. Dark wood shelves packed with books, a dark wood desk crammed with manila folders neatly arranged in rows. He sat and read my papers. Then he looked at me over narrow glasses set low on his nose. "I see here that you want to enter Nursing 208 late. Why didn't you register on time?"

"What?" I couldn't believe I was being talked to as if I were a teenager in the principal's office.

"We need to know why you're registering late," he spoke in a bored tone.

"I don't think so."

"Excuse me?" he said suddenly awake. I was answering off-script.

"Let me straighten you out. I've waited an hour and a half when I have work to do in the real world rather than being asked a stupid question by you." Barbara kicked me. I was behaving in a very un-southern manner. "I have the signature of the department chair on my add slip. I'll give you a choice: You can sign your name on that paper or we can discuss this in the Provost's office. I'm not taking any more shit from you or your staff."

We glowered at each other. He blinked and signed. Probably glad to get me out of his office before I start frothing at the mouth like one of Pasteur's rabid dogs.

The Registrar had his revenge because Nursing 208 was a nightmare. Over 70 students taking this course as a requirement before acceptance to the nursing program. With only 40 openings in the program and with the acceptances to be announced half way through the course, the stress levels in 208 were as unbearable as August heat.

To me the material in 208 was amorphous mush. After the facts, numbers and names of Pathophysiology, trying to reason through the vague nurse theorists and their inane ramblings and schemes mired me in verbal quicksand. I couldn't grasp what we were expected to learn.

And I had a larger problem. The Centrifuge plant project in Ohio had finally taken off. They needed me alternate weeks in Ohio to inventory and sell. And I wouldn't be paid for my thousands of dollars out of pocket until the sales started. Sales that I initiated.

It seems to be the rule in nursing classes that you form into groups for various projects. I discussed my dilemma with my group. I needed the money to stay in school; and I couldn't attend only half the sessions and pass.

They begged me to stay in the program. Christina was direct: "Henry, we'll take notes for you on the days you have to travel. Just don't give it up." I agreed to stay. It was one of the best decisions of my life and I wish I could find and thank Christina for her support.

When Dianne gave me the shaft (1989 de ja vu all over again) in February by demanding that I move to Ohio to coordinate sales and inventory, I could have kissed the feet of those girls. I wish I could say the same for the Nursing Department.

Two days before acceptances were announced, they called me. "Henry, we can't find any record that you took Microbiology."

"I had to take Micro before I enter the program?"

"Oh, yes, it clearly says that in the statement of entrance requirement. I don't see how we can get around this."

I wanted to yell at them: Why didn't you tell me this! But in those days I was too polite. I didn't make the list for acceptance. Later I learned that several accepted students who hadn't taken

Micro yet had been let in on the proviso that they would take it during the next semester. Why wasn't I given a similar exception? To add salt to the wound, I got a B in Nursing 208. My first sub-A grade at Kennesaw.

It would be six months before the next class was admitted. I used the time to complete the non-nursing courses including Microbiology.

In Micro during a discussion of the complications associated with *Toxoplasmosis gondii*, the professor was speaking in her usual monotone drone. "In the third trimester retinal lesions are a distinct problem."

A girl in the front row raised her hand and flicked back her hair. When called on, she asked: "Are those lesions in the mother or in the baby?"

The instructor stood in silent stun for ten seconds. "The fetus," she said quietly.

"Oh, OK." The girl buried her head in her notebook and cascading hair as she scribbled the answer furiously into her notes. Never know when that question might appear on a test. An example of a fully focused nursing student.

Nutrition was a true Mickey-Mouse course. Even the instructor knew it was wasted on us. What I most remember is a morning when I had to complete my three day nutrition assessment.

The goal of the assessment was to match one's three day intake of food with a computerized evaluation. Filling out (inventing) three days of total food intake was not the problem. Finding time to visit the college computer lab was. I chose 7:30 am on a Tuesday when I believed the lab would be empty. It certainly was, since it didn't open until 8 am.

The student-person who opened the lab could not have been any older than his pimple-faced cohorts. He appeared unfocused and bleary. I suspected that he spent the night on a typical 1990s college beer binge. No wonder we old folks consistently killed these kids on tests.

"Is this where I come to run the computer program for the Nutrition lab?" I asked.

Grunt. Cough. "Need to see your student ID card."

I took my tattered leather wallet from my pocket and searched for my ID card.

"This ID?" I asked, passing him my photo ID card.

"Yeah. But you don't have a current sticker."

"What sticker?"

"Here. On the back. See. You need a new sticker for this term. Rules state I can't give you access to the computers without a valid ID." He yawned. I was keeping him awake.

"I'm here for the Nutrition lab. Who else but a nutrition student would want to do that?" I asked. He shrugged at my question. "Are you inundated with local people trying to sneak in here to do a nutrition lab assessment?"

"Sorry, man. My central computer link can tell me that you're a student. But I can't let you in to the computer center without a valid ID. The rules are real strict."

"OK. Where do I get this sticker?"

"Student Center Information Office."

I walked down three flights of stairs and crossed the street to the Student Center. Wonderful, clear morning that I was wasting to do a stupid lab and to chase around campus to prove that I really wanted to do this lab, and that I wasn't a trespassing diet fanatic from Northeast Atlanta.

Student Information Office was closed until 8:30. I went back outside, put two quarters in the box and bought a morning paper. So much for early risers getting the worm.

I was first in line for the next showing of a bleary student. How do these kids think they're going to make it in the real world if they live with these habits now? "I need to have my ID card updated with a proper sticker."

"Yeah, this one's out of date. You need a sticker. You're a student here this semester?" he asked.

"Yes."

"Do you have your "Paid in Full" form?

"My what?"

"The printed form from the college that shows you paid your bill in full for the semester. You got it with you?"

"The semester's almost over. I couldn't be in class without being paid in full," I said in a reasonable tone. "You can pull up my status on your computer."

"I don't know about that. The rules are pretty strict. I can't give you a new sticker without that printed form."

"OK. Where do I get a printed form?"

"At the Registrar's Office." Yawn.

The Registrar again! Have I done something wrong in this life that the Dark Angel of Death follows me around in the name of the Registrar's Office?

Back outside, across campus in the other direction to the dreaded Gates of Hell. I knew they would recognize me from my last foray in the Pits and keep me waiting for hours. Would I have to sit opposite the Registrar Himself again?

I had no need to worry about that. The Registrar's Office didn't open until 9 am. Fortunately I hadn't completed the crossword puzzle in the newspaper.

Half an hour later I was in the Forbidden Zone. If the

oversized woman behind the desk recognized me, she didn't show it.

"You don't need to come in here for that. Just use the terminal in the hall next to the printer."

"Then I return here for a validation?"

"Why?"

"You mean I don't have to come back here? You mean I didn't have to wait for the past half hour?" She shook her head at me. I knew this was part of their revenge for my last encounter. My name must be flagged for perpetual harassment.

"One more question. Could I have done this from any computer terminal on campus?"

"Of course. They're interlinked."

"Even the one in the Student Center? Or the one in the Computer Lab?"

"Yes to both questions."

I returned to the Student Information Center ready to destroy the little shit. He was not there. The pretty young smiling creature in attendance added a sticker to my card and wished me a great day. I didn't have the heart to tell her what I really thought.

Back to the Computer Lab to fry that cupcake. He was gone too. The latest robot took my newly validated ID and assigned me a computer and a nutrition disc. Fifteen minutes later, I headed home.

Score: Registrar 3, Henry 0.

In the spring of 1993, the woodworking business was still doing moderately well with commercial and residential orders. The service desks and ticket booths we had built for the Georgia World Congress Center had pleased those folks so much that we were asked to do some oak display cabinets for them. Panola Mountain State Park had us construct their new dioramas. In addition, the old Morgan Building crew had reassembled at Ideal Pools, and they had asked me to work the season there. I hated pool sales only slightly less that I hated poverty. No great money in either venture, but at least I knew Barb and I weren't going down in financial flames while I continued my odyssey through Kennesaw State College School of Nursing.

In the System.

There is a truism about nursing education: When the general economy is robust, nursing schools scramble for students; when the general economy is sour, nursing schools are flooded with candidates. Luck has never stood on my right hand, and the spring of '93 was no exception. The economy turned down in '93 and when I applied for admission to the new nursing class, I was suddenly in a new ball game.

One of the differences between college in the 60s and college in the 90s is the grading system. In the 60s, the grade of A was a rare exception in a class. In Chemistry 102 (1964) I was one of two students with an average over 90. Nonetheless, the professor refused to give me an A. "You're involved in too many extra-curricular activities," he growled at me. In my Kennesaw classes there were always more As than Cs. The grading system had turned upside down.

I'm not convinced I'm any smarter now than I was in the 1960s. Nevertheless, my 3.8 average at Kennesaw was certainly more impressive than my 2.9 from Dickinson. Students applying for admission to the Spring '93 nursing class were mostly from the 1990s. They were carrying no 1960s grades as baggage.

In short: When I opened my acceptance letter, I was iced to read that I was not accepted for the nursing class. They had placed me on Standby. I was bitterly disappointed, could not believe that all my class work and hopes were flushed out to the Fulton County waste treatment plant. I was so sure of my acceptance that I had no Plan B set up. Everything was centered on acceptance into the nursing program. I had been so certain of acceptance!

Maggie was the Nursing Department secretary. I had made friends with her on the first day I visited the department. Big, sloppy, and irreverent - I actually liked her. I invited her out to lunch. And I paid. "Quit worrying, Henry. You're at the top of the list," she said between bites of a huge sandwich.

"I can't believe I didn't get in. My references were sterling." They should have been. I had sister Ann Morey (Dean of San Diego State) write one, sister Rose Ford (a professor at Rutgers) write another, and friend Lucy Gibson RN, PhD write a third.

"References meant nothing this time. They judged mainly on total GPA. They didn't want to. They had to bend to a directive from the Vice President of Academic Affairs. No choice."

"I don't understand."

"Give me a break, Henry. You're not stupid. Look at the student mix on campus. Any other decision method and the department would be open to all kinds of unprepared students. These kids are dumb enough as it is. This is not a departmental decision. But you're in. Pass me the salt."

Kennesaw State University

BSN Nursing School

Nursing 210.

Angel is a chubby, cherubic girl who lives up to her name. Her hair is bright red and frizzied out like a medieval halo. We were paired for the clinical portion of the course at a local hospital. On our first clinical day Angel and I were assigned an elderly lady on the Medical-Surgical floor.

Mrs. J was white-haired, dehydrated, aphasiac and suffering from a host of terminal conditions. "Oh, how sad," sighed Angel. The woman's middle-aged son and the physician had conferred about her condition earlier in the morning. They spoke over the unresponsive person stretched out in the bed, her eyes open but unseeing, or so it seemed. They agreed that only palliative measures were to continue. Then they left.

Angel and I stood around the room unsure what to do. We certainly didn't want to touch anything without permission. Something might break or the woman might scream out in pain and accusation. Finally our RN came in to see how we're doing.

Pat was thirtyish, thin, soft-spoken and very efficient. "Did you check the Daily Patient Care Form?"

We shook our collective heads. In fact, we hadn't the least notion what Pat was talking about. It must have been as obvious as Angel's heaving bosom. "Find her chart at the nurse's station. Look in the Daily Patient Care Form and find out what needs to be done: bath, range of motion, teeth, whatever. Do what you can and then find me. OK?"

On the form we found a note to do dental-denture care. That we could manage. Back in the room we opened her mouth and looked in. She did not react to our presence. Perfectly neat, pearly-white dentures stared back out us. The gums and mucosa were dirty with brown particles and crud. Did not look as if they had been cleaned for the whole week she had been in the hospital.

"First we need to remove the dentures and scrub them," I said. Sounded like a good nursing plan that we could manage.

We followed basic procedure and put on latex gloves. "Well, let me try," I said. I reached into her mouth and grabbed her front teeth. As I recalled, the right procedure is to pull down and out to break the denture-glue seal. At least that's what I remembered from a television commercial. I pulled. And pulled. Mrs. J groaned. That old jelly feeling in my legs returned. The denture seal wouldn't break.

Angel stepped in and tried. Pull, tug, groan.

"Got a problem?" asked Sarah, the nurse's aid who had just walked in, her arms full of clean linen. We showed her the denture sheet and described our efforts. "No problem. I've done this a thousand times. You pull and twist. Watch." She went in without gloves and pulled hard. Mrs J's eyes popped open with a huge moan. Nothing in her mouth moved.

Sarah went out in the hall and called in a loud voice: "Clara!" An elderly nurse's aid entered the room.

"What's the problem?" she asked in a clipped voice.

"We can't get the dentures out," said Sarah.

"You sure she has dentures?"

"Says so in the chart. And her mouth is pretty cruddy."

"Well, let me try." Clara reached into her apron pocket and pulled out a pair of latex gloves. She snapped them on. An old pro at this. She pulled hard. Mrs J groaned and bit down. "Yikes! She bit me!"

Pat the RN walked in. "What's going on? Am I missing a party?"

Sarah explained the situation. Pat said: "Did you pull and twist?" We nodded. "Well, let me give it a try." Pat checked the chart, put on gloves and grabbed the upper teeth.

"Watch out, Miss Pat, she bites," said Sarah.

"She won't bite me."

"Whatever. Don't say I didn't warn you."

Pat does it by the book: Tug. Pull. Groan.

Clara was massaging her finger. "Are we sure she has dentures? Doesn't feel like dentures. Has anyone checked the gum line?"

Pat pulled the upper lip back. No line. The chart was wrong. She didn't have dentures. These were her real teeth.

Later in the day I went into the patient's original folder and found the admission form. The denture information box had been checked and crossed out. On the side a note had been added: "no dentures." The transcriber missed the note and saw the check mark.

What would have happened if we had succeeded in pulling out her teeth?

Two days later Angel and I were assigned to a 36 year old man. We couldn't figure out the chart, so we sought out Professor D, our clinical instructor.

Professor D is a twin of Mary Martin in Peter Pan. Petit, perky, life-filled. "We can't find the diagnosis," I said. Angel nodded in agreement. We were an experienced team.

"That's right. There is none on the chart. John Smith is also not his real name." She stood there looking at us with a wry smile. "Come on, think it through. Look at the med orders. See this B12 shot. What's that for?"

Angel and I were as conversant as wooden cigar store Indians. I knew if I opened my mouth, it would be like walking down the hall with a gown flapping open.

"Look at the GGT lab value. SGOT, SGPT. See how high they are. What does that mean?"

It sure was nice being a nursing student for two weeks. I enjoyed it until I was shown to be the king without his clothes. And I hadn't made a Plan B. How would I explain my expulsion for idiocy? What would I say to Barbara when I got home? Or to my family and friends?

"He's an alcoholic," Professor whispered. "And the hospital has agreed not to use his real name. Do you understand? GGT is a drinking marker. Alcoholics get B12 shots. Make sense now?"

Instant relief as the disaster cloud lifted.

"Henry, draw up the B12 shot. Angel, draw up the phenegrin shot. I'll be back in a minute."

Our first injections. Ten minutes later we were still multiple-checking the dosages. Professor returned, hands on hips. "Ready?" she asked.

"Do you want to check our dosages?"

"Tell me what you did."

We explained our draws. "OK. Well, let's go in back the

room and do it."

We went in the room. "Mr. Smith," Professor announced in a strong voice, "We're here for your shots. Roll over on your left side." He obeyed her instructions.

She looked at me. "Where are you going to give the shot?"

"Vastus Lateralis?"

"Why use the leg? That's for babies. Where else?"

"Gluteus Maximus?"

"Right. Big target. You can't miss it. OK." She pulled down his pajamas.

I unsheathed the needle and poised over the rump. Spread the fingers of my free hand and measured to the right spot. Then hesitated as I reviewed steps.

"What's the wait? Do it. Stick him. Do it!" she whispered extra loud. I stuck him and he didn't scream.

"Good. Now that's over. Angel, your turn. Let's turn the other cheek, so to speak."

Nursing 310 Med-Surg.

St. Joseph's Hospital and Professor R. No more easy assignments. These were acutely ill patients that we had to care for. We had to prepare nursing plans, know the meds, and do any assigned interventions. Professor R was in charge of eight students, each assigned to one patient.

Professor R was no Peter Pan. A large woman with strong hands. A solid nurse who specialized in issues of ethics. She was fair and tough.

My first patient was AP, a 53 year old man with a history of abdominal pain who was admitted with a diagnosis of pancreatitis. He had extensive pancreatic surgery. I read the chart but didn't understand the terms.

AP was a shipping expediter originally from Queens, New York. Heavy smoker, serious drinker, no respect for hospital rules or regimen, and absolute in his views about his company.

"I made them a lot of goddamn money over the years. They damn well better put me on part time for three months when I return before I have to go back to full time fuckin' work. Bastards will probably push my ass to get right back. They can shove it.

"I've been living with this shitin' pain in my gut for three goddamned years now. First they cut me, then they torture me in

36

this fuckin' place. They don't have to put me through this crap. What do they need this goddamn middle of the night blood draw shit? Can't they tell from one bleeding stick to another what my fuckin' numbers are?"

"They need to monitor your progress," I said with lame vagueness.

"Don't give me that shit. They aren't sticking a damn needle into you every minute. And why do I need this damn IV in my arm? What's the use of that?"

I didn't know the use of that.

"Help me wheel this shit to the bathroom. Where are my cigarettes? Don't tell me I'm not supposed to smoke in there. I'll be damned before I wheel my ass all the way down the fuckin' hall just to have a smoke."

Abrasive as he was, it was refreshing to talk to a New Yorker. And I fell into his verbal style. "Listen, Mr P, if you want to smoke that shit, it's up to you. But I can't participate in you doing that kind of crap. They'll ream my ass for that."

On the whole we got along well and I was able to administer meds and keep him fairly well on his prescribed activities.

By mid-afternoon of the second day with him, Professor R called me into an empty room and firmly closed the doors. "The staff is complaining about you."

"Complaining?" I couldn't figure out what I had done wrong. "I've tried to be extra polite to everyone."

"It's your language. Henry! You're using all kinds of foul language in front of a patient. At least three nurses have come to me and said they've heard you. You have a loud voice and it carries to the hall. The loudness isn't the problem. It's the content."

"AP is from New York. That's how he talks. That's how you relate to him. It's not offensive to him. I certainly didn't mean to offend any of the staff."

"This has to stop or you'll be removed from the floor and I'll have to report you."

"Yes, professor, I'll stop." Visions of Neotronics and job loss. I was flooded with feelings of failure and insecurity.

CRH was an obese 44 year old white male with a history of diverticulitis and admitted with severe abdominal pain. He had a subsequent sigmoid anastomosis (cutting out part of his colon and rejoining the rest) followed by peritonitis (an abdominal infection) and an open abdominal incision that went from pubis to xyphoid.

CRH had been a management trainee at a local company, a job that he claimed he enjoyed. After admission to the hospital, his employer terminated his employment and informed CRH that he had no medical coverage since he had not reached full-time status.

His wife was also unemployed. They had two children. They expected to lose their savings and their home as a result of this hospitalization. CRH was a "charity" patient at the hospital.

"I know God has a purpose in this," CRH told me. "I know we'll come out stronger." His eyes had a vacancy that belied his words.

CRH probably belonged in the Intensive Care Unit (ICU). He had a nasal-gastric (NG) tube sticking out of his nose, Oxygen at 4 liters per minute by nasal cannula, a central line that came out of his left chest for total tube feeding (TPN), two peripheral IV lines, one in each arm, for fluids (D5W), antibiotic piggy backs on the IV lines, and a pain medicine (PCA) pump. He also had a colostomy pouch for fecal matter until his intestines healed and a Foley catheter in his penis for urine collection.

My lingering memory of CRH is the stench of fat when I first pulled fouled bandages from his open wound to do a wet pack. His incision was a wide canyon with at least four inches of exposed fat on both sides. At the bottom of the canyon was an open passage to his peritoneum. My mind went back to my 1970s farm in rural New York State and pig slaughtering time.

During my second day with CRH, he became confused, his temperature spiked, his pulse increased and his blood pressure fell twenty points. For the full day I worked as hard as I could with the

RN and the physician. Professor R was in and out of the room a dozen times an hour, watching over me like a buzzard waiting for the last flicker of life.

Several days later I met her at school in her office.

"You weren't pleased with my work with CRH?" I asked in my usual round-about way of opening a conversation.

"Quite the contrary," she answered with surprise in her voice. "I think you should know that your constant nursing probably helped to save his life. He was on the way out."

Nursing 311 OB-GYN

I observed one day with Ellen, a mid wife at a private OB-GYN center. She was a small, somewhat dowdy lady in her early fifties. She showed me pictures of her teenage children and her live-in female partner. She left me to my own conclusions.

Our first case was a 35 year old black female who had been pregnant three times with no live deliveries and three abortions. She presented with the inability to conceive with her present partner of the past year. She wanted to become pregnant and have a successful pregnancy with a healthy baby.

Fifteen years previous she had two abortions and an ecoptic pregnancy that had led to the removal of her right falopian tube. She talked to Ellen about her fears and pushed hard for us to give her a dye test to establish the patency of her remaining tube.

I cut into the conversation: "Has your partner had a sperm count?"

"Nnno," she said hesitantly.

"Does he have any children?" I pressed forward.

"None that I know about. He says he suspects he's made some women pregnant in the past."

"A sperm count seems to me a simpler procedure than a dye test. Why do we assume that the problem is in the woman?"

Ellen jumped back into the conversation. Maybe I had pressed too far. Maybe I had stepped out of line as a student. She asked the lady to keep a temperature chart for a month and gave detailed instructions on the procedure.

"We're going to arrange that dye test for you next month. And I want you to take this cup back home. Have your man do a sperm sample and bring it back in here first thing one morning next week. Then two days later you can call for the results. We're going to help you have your baby."

Later in the morning in the hall we read the chart of a 52 year old divorced black female who came in complaining of vaginal itching and discharge. We entered the room and found a stylishly dressed, professional lady.

"It's a Tric infection," she said before Ellen could even make introductions.

"How do you know it's trichomonas?" asked my midwife in a calm voice.

"Happens to me all the time. The itching. The awful smell. They always give me Flagel and the Tric goes away for a couple of months. Then it always comes back like a bad penny."

"When was the last time you had sex?" asks Ellen.

"The last time was December 18," she answered without hesitation.

"You sound pretty definite about the date."

"When you only have sex twice a year, you remember the date." We all laughed. It was a cute line, but I didn't believe her.

I sat on a chair behind the patient so as not to cause embarrassment during the physical. Ellen dilated and scraped and smeared a slide. "I'll be right back after I look at the slide."

In the hall she said to me: "I'll place bets it's a bacterial infection and not Tric. I don't hardly have to look in the microscope. Know why? Trichomonas is always associated with sex and she's not active enough."

On the slide we both saw dimpled, scrunched epithelial cells. No flagellated parasites that look like bouncing Bettys. "OK," said Ellen. "Here's your nursing question. The treatment for the bacterial infection or Tric is the same. Do we tell her it's bacterial vaginosis? Do we correct her self-diagnosis or let it stand?"

"She's so nervous that it might be better to let her own diagnosis stand. But that's not what I would do. We should tell her the truth. Otherwise some day someone else will tell her the real score. Then she'll either call that person a liar or think of us as incompetents."

Ellen smiled and took out the prescription pad. "You may make it yet."

Jane was a 46 year old white female who came in for a basic

43

yearly exam and consultation. "I haven't been sexually active for almost two years. Recently I met a man I really like. Things are going great with us and I want to be sexually active with him. Please don't think this is a dumb question, but what kind of prophylactics should I insist that he wear?"

"What kind of prophylactic?" Ellen asked with an unsure note in her voice.

"Well I've heard about lubricated and unlubricated, smooth and ribbed. I just don't know enough."

Ellen looked right at me, straight into my eyes.

"May I answer that?" I said smugly. "Here's what you need to buy and have handy..."

Later, in the hall, Ellen thanked me. "That's not a subject I know a lot about, as you may have guessed, with my life style."

I fell in love with gynecology that day. There was so much teaching and interplay. Your patient wasn't half dead in a bed with tubes in every orifice. Ellen told me that Georgia had a few male midwives, but it was a difficult profession for a man. "The patients will accept you. You have a nice way about you. Unfortunately the female nurses will subject you to an awful discrimination. They're very catty."

Independent Research.

Nursing is a succession of courses in lock-step order. You are only allowed to choose two electives from a group of about five. Ethics 420 was one course I wanted to take. The others struck me as Mickey Mouse.

In 1987 when I was developing an incontinent pad for a medical soft goods company, I had an idea for a one-part pad using reticulated foam for the center soaker instead of the usual cotton/poly felt. The top would be brushed poly and the backing a light weight urethane. The foam would allow for the three pieces of the pad to be flame-bonded together which would be a marked advantage over current products that are separate pieces sewn at the edges. Light and strong with good absorbency. A sure winner. This foam pad idea stayed with me for years in the back of my mind.

In late 1993 Barb and I attended The Bobbin Show in Atlanta looking for new sources of supply for the hotel bed straps (another small business we had started to generate cash during my school years). On a back aisle, in a small booth, was a man from Commerce, Georgia who did custom flame bonding with foam. His major product was headliners for autos and trucks. And he was willing to listen to a crazy idea from a nursing student.

A few weeks later he produced a small run of material that Barb and I sewed into incontinent pads. Next we needed to arrange for a local nursing home to do a trial of the pads. And concurrently I needed to fulfill one of my elective requirements. Why not propose an independent study project using this trial? I spoke to Professor B.

"It might be possible," he said. "Put your ideas on paper. We will certainly have to submit to the Institution Review Board for approval."

"Why?" I asked smelling my first problem.

"You always submit to IRB when you do research using patients. And while I think of it, talk to AW about the nursing homes. She works with them all the time."

I spoke with Professor W who suggested I call on a local nursing home and speak with Connie the DON who was a friend of hers. She also agreed to act as my professor/coordinator for the independent study.

Connie was most welcome to the idea of trying (at no cost) a dozen pads. That night I spent hours constructing a proposal for Independent Research. That night was also the last day to register for electives. No worries. I had a solid proposal and a good idea.

I submitted the proposal to Professor B who forwarded it to the IRB. He called me two days later. "They want a signed consent form from all the patients who will be using the pads."

"It doesn't work that way. Pads aren't assigned to individual patients. They're assigned to a wing or a floor. And how do I get consent forms signed by patients who aren't cognitively capable."

"You may have to go to the families in that case."

I thanked my BSN Department Chair politely, hung up the phone, and turned the walls blue with curses. I called the Registrar's Office (Not Again! Please, Lord, have some pity! I'll be good.) and learned that Ethics 420 was full. I also learned that Professor R (from Med-Surg) was the instructor. I faxed her a note. I phoned her. I begged and groveled. She let me in.

Barb and I proceeded with the incontinent pad trials at the nursing home on our own. The pad was a winner in every department. Residents said it was the warmest, most comfortable pad they had ever used. The laundry said it was easy to wash and dry. The nursing staff said it was light and fluffy, and it gave them a sense they were adding comfort to their patients.

We decided to go to the Georgia Nursing Home Association meeting later in the year. We bought a surger to do the edges of the pad and ordered 500 yards of material from the flame bonder. That's when the supplier notified us that the price had doubled from $4.75 per yard to $9.50 per yard. Each yard yields two pads.

At the lower price we had a product competitive price-wise

with the high end of the incontinent pad market (about $119 per dozen). At $9.50 per yard we were priced out of the market. End of product.

Nursing 312, Pediatrics.

The feared Professor K. Every upperclassman I spoke to said the same thing: "Stay away from her. A good grade from her is as rare as a Kennesaw virgin." Professor K taught pediatrics at the Atlanta Fulton County Hospital: Grady Memorial.

Professor B came into 311 near the end of the semester. "So many of you want to do Peds 312 at Scottish Rite that we are going to run a lottery draw to be fair on assignments around town. You will each pick a number out of this hat. Low numbers have first choice." Our class now had 62 students. I picked number 57. That's how I ended up at Grady with Professor K.

K is a synonym for tough. About my age, army nurse background, fiercely competitive tennis player and runner. The world in black and white.

One morning I watched her approach a young female student who was drawing up a med. "What med are you preparing?" K asked, her hands firmly planted on her hips.

"Nafcillin," the student answered.

"What's the maximum dosage for a child aged 8?"

"Maximum dosage?" The girl's voice held a slight wobble of uncertainty.

"Yes, that's what I want to know. What's the maximum dose in milligram per kilogram for your child, age 8."

"Uh..." I watched sweat form on her forehead.

"You don't know do you? And what are the side effects?" K's voice hadn't raised a decibel, but the power was streaming forth like red lava from an Icelandic volcano.

"Uh..." The student's hands started to shake.

"Did you read about this drug last night? If that were your child in there, would you give a drug you didn't know about? Not to my kid you won't. Get your coat and go home. Next time don't come to my clinical, or any clinical, without knowing your drugs."

The girl started to cry. "I'm sorry, Professor. It was so late last night..."

"I don't want to hear it. There is no excuse. Leave now." K turned her searchlight on me. "What drug are you drawing up?"

I told her: Nafcillin.

"Uh-huh. Another student with Nafcillin. What is the maximum dosage for a child?"

"200 mg/kg IV per day," I said with confidence. Us old folks do our homework.

"Hmm. What are the side effects?" Her eyes stared firmly into mine. The snake was not letting the mouse run away.

I told her what I remembered: "Vein irritation, deep vein thrombosis (DVT), and nausea and vomiting (N&V)."

"Anything else?"

"Not that I recall."

"You're missing an important side effect," she said.

"Then I've forgotten it. I'll look up the drug again before I deliver it so I don't harm the patient."

She nodded her head in philosophical agreement but didn't let me escape. "What about the Stephens-Johnson syndrome?"

I hold my breath and jumped off the balcony into empty space. "It wasn't mentioned in the medication book."

"Not in the Nurse's Manual! Impossible! *Get me the book!*"

I brought her my book and thought how much fun it's been in the nursing program. Then a gift from the gods: The syndrome wasn't in the book under Nafcillin.

"Ridiculous! How could they omit that? *Bring me the Nursing PDR.*"

I reached up into the cabinet and brought out the PDR. She took it from me and leafed through the pages. On the page for

Nafcillin she read and let out a snort. It was not in the PDR. Could she be wrong? How would she extricate herself from this?

"What kind of a PDR is this? The charge nurse has a pediatric drug book at her station. *Bring that to me.*"

K flipped through the book snapping the pages one at a time. A broad smile formed. "There it is! Stephens-Johnson syndrome. Henry, learn it for next time." She slammed the book down and walked on to the next student.

"Yes, professor," I spoke to her back.

K taught me the *RN Attitude*: When you know you're right, you stick to your guns. When you see someone doing something potentially damaging to a patient, you stop them. When you do something wrong, you admit it and then state how you are going to correct the problem from the point of view of patient care.

I had a dream that night. Professor K was lecturing on Japanese nursing interventions. Around me students were furiously scribbling notes. I found two oak boards, set them between two desks and went to sleep. When I awoke, everyone was off in various rooms engaged in procedures based on the lecture. Complicated, invasive procedures that nurses should never be performing. Everyone was talking in rapid medical jargon like interns conversing in a hallway. I could not follow the talk and felt completely inadequate.

A solitary nurse walked down the hall towards me. She carried a huge basket of shucked ear corn. She tripped and dropped the basket a few feet from me. I helped her put the ears back in, and we carried the full basket down the hall to a room set aside for farrowing pigs.

Half a dozen durocs and piglets were scrambling around in piles of filth and manure. We threw them the corn which they fought over and ate with great grunts of pleasure. I thought to myself: This is the life I love; I can understand animals better than nursing procedures.

Oxymonias (Oxy) was a five year old ball-of-fire black boy with severe sickle cell pathology. In the hospital for an infected portacath (from a fall), seizures and strokes secondary to his sickle cell. On the face of it, a dismal situation with a poor prognosis, except the boy was a running, jumping smile, a constant stretched IV line and bright saucer eyes that followed me everywhere. In the room with him was an older, very large black woman who seemed to be asleep in the Lay-Z-Boy lounger.

"Hey, what's your name," he said as soon as I entered the room.

"My name is Henry."

"My name is Oxymonias. Are you a doctor?" he said bouncing from one foot to the other.

"I'm a nurse."

"You a nurse? You're not a girl."

"No, I'm a man nurse."

"You going to listen to my chest?"

"Yes."

"Can I listen to yours?"

"Sure."

"Now?" He grabbed the stethoscope; I opened my shirt and put the stethoscope bell on my chest and the ear plugs in his ears. He listened and smiled.

"You going to look in my ears with that light?" he asked

"Yes."

"Can I look in yours first?"

Later I spoke to the woman in simple terms since I judged she probably wasn't well educated. "I have to give Oxy a strong penicillin-type medication that has a particularly nasty side effect. It doesn't happen in every patient, just a few, and I need to explain to you…"

"Mister, you talkin' 'bouts Stephens-Johnson Syndrome? He don't gets that."

So much for condescending talk to stupid black women.

Around midmorning I noticed that before I could set up the piggy-back Nafcillin, I had to change the tubing lines because it was out of date by 24 hours. Welcome to quality nursing at Grady. I had begun that morning sweeping out bloody bandages and used hypodermics from under his bed.

From class I knew the best way to change tubing was to prepare the IV piggy-back bag in the med room behind the nurse's station and to set up the new tubing in the patient's room. At least that's what I remembered from 310 Adult Med Surg.

I located the new "spaghetti" drip tubing and went back to Oxy's room to redo his piggy-back line that empties into his regular IV line. I certainly wasn't going to change his main line or mess with his insertion point in his arm. Let the real RN do that.

I walked into Oxy's room with the tubing box right in front of me. His sunny face folded into fear and panic.

Stupid! I screamed inside myself.

"It's all right, Oxy. This is not for your arm. This is for the bag. No needles. I promise you. This is only new tubing for the bag. To make it safe for you. No needles. That's a promise." He looked at me with suspicion for the first time that day.

I hid the incident from Professor K. Partly from fear of her rebuke, and also because I didn't need to add to the burden of her day when I knew the parameters of my action and how to correct those errors.

Needless to say, I left Oxy's room with the tubing in the

pocket of my white jacket. Back in the medication room behind the nurses' station I attached the spaghetti to the piggy-back bag. Where I should have done it in the first place.

Oxy's bright ebullience lulled me into forgetting that he was a little boy in the hospital possessed of the fears of bodily invasion that go with a five year old. I acted like a visiting friend of the family and not like an RN assigned to care for this boy. That was a tough way to learn that there are times when distance is critical to the patient's care and well-being.

That was the bottom line. Oxy was in the hospital because he needed competent medical intervention beyond what he could receive at home. My role as a nurse was to help him achieve optimum health so that he could delight in the natural energy and joyfulness that were at the core of his nature. When I became sloppy and failed to view my patient as a sick little boy, I failed in my duty and brought pain to that boy.

I feared pediatrics because it colored my transgressions in neon and led me down paths of extreme self-criticism. K once told me: "95% of the time what you do won't cause harm. The other 5% of the time I'm here to back you up." More than anything else what I wanted from pediatrics was to finish the course without harming a child.

One day I chose an AIDS baby for my patient. Technically the two year old black child could not be classified as having the AIDS syndrome yet. Jozelle required a second major opportunistic infection to receive that official label. But why be so technical? She was born with the human immuno virus, this was her first hospitalization for a major disease state (candidias from mouth to anus), and her future was nasty and brutish.

Both parents were HIV positive. I entered the situation with a head full of preconceived notions and stereotypes. To begin with, AIDS is a charged subject with me. The best friend I ever had, the only person I have ever been able to fully open to, died of AIDS in 1986. A horrible, fistula-filled death for a man who delighted in his physical robustness and who was fastidious in his personal appearance.

In addition, my youngest brother died from AIDS. He was an IV drug user who watched his friends die of the disease one by one. When Paul's teeth began to fall out as the infections started their lethal cascades, he filled a syringe with a designer drug and left the world ensconced in the arms of his greatest love.

This was some of the baggage I brought to Grady and my small patient. I entered her room with visions of an orphan child whose parents were severely ill, lost in the woods of their own illnesses, unable to care for their daughter. Unmarried blacks making an unwanted baby doomed before birth to a miserable, short existence as medically painful as their own. White, suburban stereotypes from a man who boasts of his varied cultural experience.

In fact Jozelle was out of bed in the rocking chair being spoon fed and talked to by her mother. Jozelle's mother was withdrawn, had poor teeth, deficient hygiene, and lacked the strength to provide Jozelle with enough play stimulation. But she was definitely well bonded and very loving towards her daughter.

At 8 am the baby's diaper had already been changed by the mother and the Nystatin cream liberally applied. The sound of shattering glass was my own polished images falling to the floor.

Around noon, a black couple entered the room. The woman in a great flowered dress, the man in a huge floppy hat. Jozelle's mother introduced me to the little girl's aunt and father. We shook hands in a limp, affected greeting.

The sisters went down to McDonalds while the father took Jozelle out of the crib to the rocking chair. He kissed her, held her close and fed her half a bottle. Another glass idol fell to the floor, a splattered stereotype.

Welcome to Grady Memorial Hospital.

The next day I spent with Darvin, an acutely ill 20 month old boy. I did not believe he would live to see his second birthday. My second small child for whom life promised a short, unhappy existence.

More than the dismal prognosis, I was bothered by his unrousable lethargy. No toys in the crib, no visual or auditory stimulation. His mother never came to visit in the two days I had him as a patient. His only tactile stimulation was from the nurses.

He was in the lowest 5% for weight and head circumference but at the 33rd percentile for height. A mismatch that shouted out poor nutrition, growth and physical development.

During the morning, in an effort to clear his lungs of congestion (oh the wheezing in the stethoscope!), I sat with him in the chair. I thumped his back with a cupped palm to break up the congestion, his head thrown across my shoulder. How many nights had I held my own sick kids like this? How easily the feelings and memories of 20 years ago can be summoned.

I talked to Darvin about my problems and concerns in school, about how I hadn't studied for the midterm yet, about my marriage and business problems, money tight and wife away at work too late for too many nights. His head hung limp against my shoulder, a putty rag doll. It appeared to me that he listened to me as well as the last two shrinks I had spoken with.

K swooped in the room and gave me a stern look. "Henry, what are you doing?"

I gave her an equally stern look back. "Darvin and I are rocking in this chair. We're having a long talk about our problems."

She stared at me for a moment, unaccustomed to that tone of voice from a student. Then she said: "OK. Sometimes that helps." And she left as quickly as she had appeared.

I came into the nursing program thinking that I was going to get my RN, my magic piece of paper and return to the business world more fire-proof. Instead, here I was rocking a sick child, behaving like a little nursey-nursey with a starched white cap. And it felt right.

Later in the day K returned. She too was concerned about the child's unbreakable lethargy. While we were discussing the case, a young woman resident with tired eyes entered the room and gave the child a quick check.

"This child is too lethargic," K said. "We need to do a CBC and check RBC and MCH."

"I'm not sure that's advisable," answered the resident. Mistake.

"You don't? Well, doctor, in your examination did you notice..." K rattled off five points that left my head spinning. I wasn't the only person in the room who was impressed and persuaded.

"I'll write the orders," answered the young resident as she left with softer steps than she entered.

Professor K and the RN attitude.

David and I were assigned as student observers to the Pediatric Urgent Care Department for one day. David is a big, very big, nursing student with a mild disposition hiding behind a body-builders physique. We worked the morning in triage with Chris, an RN from England who had come to America to study our system and procedures. Chris has a fair complexion, flaxen hair, and a soft, comforting voice.

Midmorning, a striking black lady with a bouffant hairdo that made her look like a romantically inclined rooster came to us with her young son in tow. A huge barge pulling a small sail boat. The boy balked at every step; his mother alternately pulled and prodded him along. He looked to me more terrified than recalcitrant.

Chris asked the boy to step on the scale so he could get a weight. The boy balked and shied away. The more the mother pushed him forward, the more he leaned back. Finally Chris had mother and son get on the scale together. Then he had the mother get on alone. Subtraction gave us the weight of the boy.

The boy had a slight rash on his arms and chest. Chris asked a series of questions in a soft tone to the boy. Each time the mother answered without giving the boy a chance to say a word. Chris had the patience of a Buddha.

Finally, the mother said to Chris: "Doctor, there's something else..."

"No, ma'm," he interrupted her with his bright English accent, "I'm a nurse."

She stiffened and transferred her gaze to me. "I'm a nurse, too," I said to her unasked question.

She looked at David who is nearly twice my size. "We're all nurses," he said.

"You wanted to tell me something?" English Chris continued in nearly hushed tones.

"I'll tell the doctor."

"If you tell me, I can write it in the chart for the doctor to think about before he sees you."

"I'll wait and tell the doctor. I don't want to talk to a nurse."

"That's fine," Chris responded smoothly. "Is your son allergic to anything?" He continued his understated questioning as if the doctor-nurse exchange had never happened. He was unruffled by the woman's stiffness.

"Penicillin and pork," she answered

"Pork?" I asked. "How does he react to pork?"

"We're Moslems. I ate pork once and I don't want my

children to have the experience. My daughters almost ate pork once. I don't want the same thing to happen to my son."

Chris signed the various Grady forms and directed the woman to a waiting room. I gave the boy a smiley sticker for holding still during Chris's short examination. The boy smiled for the first time. Then his mother gathered him in her great dress and set off down the hall in full sail to her assigned waiting room.

Chris, Dave and I looked at each other and shrugged our shoulders. I wasn't sure what more we could have done to encourage the mother to talk to us about her son's condition.

Like so many people, she believed that medical cure resides in the physician, and nurses are functionaries reaching beyond their grasps. And male nurses must be rejected medical students or otherwise impaired intellectually. Well, there are times when I wonder about that myself, as it applies to me.

Or that may be pushing my insecurities too far. Maybe she was stressed to begin with. The comments about pork had a good dose of inappropriate reasoning. The exchange with Chris about not sharing information with a nurse gave her the opportunity to be in control, to feel some degree of power over three white non-Moslem men who were the gate keepers to her son's medical treatment.

Stereotypes, racial delusions, reflected paranoia: the psychological battleground of Grady Hospital.

CB was a 17 year old black male who presented to Grady with a complaint of epigastric pain for two weeks duration. He had a history of sickle cell disease with multiple hospitalizations and a gall bladder attack six years previous. Three months before I saw him in April, he had been admitted to Grady for abdominal (GI) vaso-occlusive crisis related to his sickle cell and complicated with duodenal ulcers.

He took Prilosec and folic acid for his ulcers. During his January admission he had been treated with Amoxicillin for his bacterial infection and Nubain for pain associated with his sickle cell pathology.

I went down to Grady on the afternoon after his admission to pick out a patient for the next day's clinical. We were required to locate a patient, review his chart, prepare a nursing plan, and learn about the drugs we would be administering the next day. This entailed a long trip downtown, a long trip back to Roswell and the prospect of a long evening writing and studying.

My nursing clinical at Grady meant that Barb and I were both downtown. Car pooling sounded like a great idea, but her schedule at Marriott engineering was so erratic that we couldn't find a way to carpool.

I met David in the hall. He was also searching for a patient. "Hey, Henry, why don't you take CB? I talked to him earlier. He's had student nurses before. He's an interesting teenager with sickle cell and pain. You could use a break from these AIDS patients you've been focusing on."

After reviewing CB's chart, I went into his room and introduced myself. Besides his obvious shyness, he impressed me as intelligent and knowledgeable about his condition. We discussed his medical history and present treatments in general terms. I needed background information to help write his care plan.

He repeatedly emphasized the constancy of his abdominal pain. "It hardly ever disappears." In his chart I read that he was taking Nubain 20 milligrams every 4 hours through an IV line. Even without checking my medication book, I knew that was a large dose of narcotic for a boy weighing 115 pounds.

Morning report begins at 6:45 at Grady. As students, we needed to arrive at least 30 minutes earlier to check the medication book for any changes and updates. It was our last opportunity to cross-check new meds with the med guides before Professor K appeared and started asking questions. And I knew where that could lead. On CB's med sheet that morning I learned that his

Nubain had been changed to Morphine 10mg Q 4 hours IV. That dose seemed more in line with his age, weight and reported pain.

Morning report was in the cramped break room which apparently was off limits to the night cleaning crew. Trash, dust and a stained table top constituted the room's natural condition. Of course, accumulated detritus didn't seem to motivate the cleaning staff in a patient's room either.

Room by room the night nurse mumbled her way through the patients. She was talking to eye-lid drooping day nurses clutching steaming Styrofoam cups of coffee.

"Room 216, bed B. CB is back for his monthly fix."

"Nubain again, I bet," mumbled a thin black nurse.

"Changed to Morphine 10mg last night by the resident." Laughter filled the room. "Another new resident in the learning curve." Chuckles and head shaking.

"Boy got what he wants again," said an older white nurse with ill-combed hair. "Wonder what pain story he used on the doc." She sipped her coffee. "He got a student nurse today?"

I raised my hand.

"Good luck," she said raising her coffee cup in salute.

My first thought was that I was viewing inappropriate and unprofessional nursing in both judgment and expression. My second thought was a worry that maybe I had misjudged CB yesterday. Maybe these experienced nurses knew something I had missed. Had I been manipulated by the patient?

In my first physical assessment of CB after report, I asked him about his pain. "On a scale of one to ten where one is ethereal bliss and ten is the devil's hell, where is your pain?"

"Ten," he answered. He spoke calmly but his eyes were shining.

"Where is this grade ten pain?"

He pointed to a spot in the upper epigastric, the same spot he had shown me yesterday.

"Describe the pain. Is it diffuse or localized?"

"Local, intense and focused like a fist."

He certainly knew how to answer the questions in the right order. I knew from David this was not the first time he had a student nurse.

"You've been with this pain for over a week now. You've taken Nubain and now Morphine. What medication works best to relieve the pain?"

"The first medication they gave me: Amoxicillin."

That was a strange answer. My internal eyebrows raised a notch. Amoxicillin is an antibiotic not an analgesic. CB had to be mistaken. Why invent a sharp, localized pain that you claim is best relieved by an antibiotic? A diffuse, deep pain would be a better choice to influence a physician to prescribe narcotics for relief, if he was hooked on the drugs as the nurses had so sarcastically remarked.

Later in the morning, out in the hallway, four medical students and their professor were doing rounds. I joined them. One student young enough to be my son asked the question I had been thinking: "Why would sickle cell pain be so localized?"

The MD professor was a large man dressed in a herring bone jacket and a yellow bow tie. "Pain is person," he pontificated. "The pain may not be directly related to the vaso-occlusive pathology. Yesterday he told me that Amoxicillin relieved his pain best. I ordered a gram-positive antigen test that should be back today."

"It was drawn at 4:30 this morning," a med student commented. "It won't be ready yet. Possibly by 4 this afternoon."

"Possible but don't bet the store on it. 18 hours from the original order to the blood draw. Not a record, but close. Well, maybe we ought to consider a return to the antibiotic regimen. Curious case. 10mg of Morphine should only provide half the relief of 20mg of Nubain, yet there seems to be no difference in reported pain relief. Neither medication appears to reduce the pain completely."

During the day I had noticed a cloth covered Bible next to his bed. "Is this yours?" I asked.

"It belonged to my papa. He gived it to me. I read from it every day. Religion be my rock."

We had spoken at length about his participation in the JROTC military program at his high school. It was a source of pride and trouble to him. Pride in his accomplishment and trouble when his sickle cell flared up if he pushed himself too far physically.

CB knew the powerful analgesics relieved his pain and he was glad for the relief. But he was determined not to scream, cry or complain about the pain. "This pain be from God. Don't make no sense to complain about something from God. Just accepts it as best you cans."

At the end of the day I was firmly of the opinion that the morning nurses were wrong in their assessment of CB. They had missed the pain because CB is deeply religious and had a fine-tuned "macho" self-image.

Two weeks after my day with CB, I met the medical professor in the pediatric ER room at Grady. He wore the same jacket but sported a white and blue polka dot tie this time. "Professor, I wanted to know if you recall a teen-aged black boy named CB who was in the hospital with sickle cell and upper epigastric pain."

"I remember the boy well."

"May I ask what happened with the case? I was his student nurse for the day."

"I remember you, too. An odd nurse that joins the medical rounds. Well, strange thing about that case. The gram positive antigen test came back non-reactive. Then over the next several days we altered his morphine levels without telling him. His description of his pain site remained the same. When I looked deeper into his history - and he has a considerable history with this institution - I found that there had been earlier suspicions of inappropriate narcotics use. I came to the conclusion – quite reluctantly I might add – that he's hooked on the drugs. He fooled us for a while. That's not very common, I hope."

This case is a paradigm of mismatched viewpoints. Vaso-occlusive episodes secondary to sickle cell pathology will be a life-long problem for CB. The hospitalization that I participated in was not his first and will not be his last. He was an informed, educated teenage patent, experienced in hospital procedures, personnel, medication policies and the various pain antagonists. He presented the health care providers with a pain scenario that was sufficiently outside the norm that suspicions were allayed.

The staff nurses who had experience with CB from his earlier admissions were firm and sarcastic in their beliefs that CB's need for narcotics was driven by his "needs" and not by actual pain.

They administered the prescribed medications as ordered. As far as I could tell, they did not approach the interns or the residents with their concerns that CB was a junkie manipulating the system.

The physicians treated CB according to established protocol. They did not seek out the nursing staff to inquire about his past hospitalizations and narcotic usage even when they had questions. The professor sought enlightenment in the medical chart not from the people closest to CB's care.

The various medical groups acted independently, each group cloistered with its peers. This closure resulted in CB's successful use of the system to obtain his narcotics. It also reflects poorly on the overall quality of care delivered to CB. There has to be a better way.

Ethics 420.

By the spring of 1994 the hotel supply business with those little straps and door stops that Barb invented was doing so well that I was able to retire the woodworking shop. Lev, the Russian immigrant from down the street, agreed to store my woodworking tools in his garage in exchange for his occasional use of them. That freed up space for me to expand the manufacture of the hotel products. Sewing was brain dead boring, clean, safe and profitable. Goodbye saw blades, hello sewing machines.

Ethics was a required course. Professor R again. Since we were out of the hospital, I didn't worry about being called on the carpet for inappropriate language. The main thrust of the course was to discuss issues we had encountered in our clinical experience that involved ethical dilemmas. I chose an OB case from Grady that greatly bothered me at the time.

AB was a 33 year old black female who was in Grady for pre-term labor along with severe puritis (itching) secondary to gall bladder disease. She was 28 weeks pregnant with twins, each estimated at 525 grams (13 ounces). She came to Grady with a demand to have the babies delivered so that she could end her pain and discomfort.

AB was a small woman, slight of build. She had a history of sexual abuse as a child, physical abuse in her current relationship (, multiple active sexually transmitted diseases, positive tests on admission for alcohol and cocaine (in spite of repeated and vociferous denials that she used either during her pregnancy), two elective abortions in the past two years, a heroin overdose suicide attempt two years ago, and two arrests for disorderly conduct in the past two years. Not exactly a picture of an alluring Southern belle. Welcome to Grady.

The staff wanted to start MgSO4 (magnesium sulfate) therapy by IV drip to hold back contractions and delay labor. AB's primary care RN was a black lady with a no-nonsense demeanor. We spoke at the nurses' station away from AB's room. "If we can delay delivery until those babies are 32 weeks, then they'll have an improved chance of survival. The key thing is their lung development will be better. 30 weeks is our minimum goal. At 28 weeks and 525 grams the babies – if they survived the cocaine and alcohol withdrawal – will spend months in the neonatal intensive care unit at a huge cost in hospital money and resources. We can avoid this with a two week extension in gestation without the substance abuse."

"So it's cheaper to keep her here for two weeks than to keep two babies in ICU?"

"That's the game, Henry."

When we presented AB with the need to start MgSO4 therapy to protect her babies, she refused. "You brings me in here and tells me what to do. Do this, do that. Nobody listen to me; nobody ax me what I wants. I ain't doin' nothin' till I gets me a good dinner."

The nurse spoke quietly. "AB, you want to have these babies healthy. I'm a mama too. We need to stop these contractions, to stop this pre-term labor, to stop this pain. You need to have this medicine. And we can't do that unless you eat nothing by mouth. Those are the doctor's orders. You'll be fed through the same IV line that delivers the magnesium sulfate medicine. You want your babies healthy, don't you?"

"Yes, I do. I wants these babies. That's why I'm here. Now I'm hungry. I ain't been fed all days. You brings me dinner, I lets you put that stuff in my veins."

Several more times the RN and other staff members including an intern discussed with AB about the best way in infuse and monitor MgSO4. She was steadfastly adamant: no food, no medicine. In the end AB wore them down. And she insisted on food that she wanted: a Big Mac with French fries.

Later in the day AB told me that she never wanted these babies. They certainly weren't planned. She just never got around to aborting them the way she had with her last two pregnancies.

Two weeks into her hospital stay, AB announced forcefully that she wanted to end her discomfort by having the babies delivered. The staff opposed this because at 29 weeks the preemies would be in extreme danger. Much of AB's discomfort could be relieved by treating her gall bladder dysfunction, but that treatment was contraindicated by the advanced pregnancy.

Once again the staff had to bargain with AB. She agreed to tolerate the MgSO4 treatment for another 14 days provided the hospital would deliver the babies at that time. "Else I pulls out these here tubes and goes home."

AB had minimal family support. During her stay in the hospital, the boyfriend and her children never once came to visit her. Her mother came for a single afternoon. She sat in a chair on the other side of the hospital room clutching her purse in front of her obese stomach. She never touched her daughter and she spoke to me with detachment.

"I don't like to see my daughter like this. This is distress on her. Bad distress from God. This be God's punishment for her lifestyle. Yes, it be."

From an ethical standpoint, this was a complicated case because the participants had differing agendas. From AB's point of view, these were unwanted babies that she neglected to abort. She came to the hospital in pre-term labor expecting to have the babies delivered. She certainly did not plan on a multi-week unpleasant stay in the hospital. She knew that the medical personnel were not treating her as the primary patient – that position was held by her unborn, unwanted babies. She told me that her boy friend claimed that the babies weren't his and he didn't want to be part of their lives or care.

From the physicians' point of view, allowing AB to deliver the babies when she was admitted at 28 weeks had the potential for substantial harm to the fetuses including death. They also knew that providing the best care for the fetuses brought AB into danger with her untreatable gall bladder disease and MgSO4 treatment.

From the nurses' point of view, they wanted to bring the pregnancy as far into term as possible to improve the normal growth and development of the babies in utero. Additionally they knew that pre-term babies needing extra care are not a good match to a mother living in an abusive household and addicted to multiple substances. AB's family coping skills with her current three children could only be compromised by the twins.

The nurses had a conflict over who was their client. Was it AB, or the babies, or the family unit? Because AB was distancing herself from her unborn babies, advocacy for AB was not the same as advocacy for the babies.

The nurses agreed to bargain with AB on her therapeutic regimen in order to be care givers to both her and her babies. The result of this bargaining was to reduce the efficacy of AB's treatment, to open the door to litigation based on compromised treatment, and to allow AB to become the determinant of her treatment based on her manipulation of the system. However, the nurses knew that without bargaining into a compromised treatment plan, AB would have refused the MgSO4 and delivered the babies into great danger.

From the hospital's point of view, they wanted the babies kept in utero until 32 weeks in order to reduce the time and cost required in the Neonatal Intensive Care Unit (NICU). Below 30 weeks the babies would be in NICU at least two months. Every day on the ward in utero was ten times less expensive than babies brought into NICU. Once AB had been accepted as a patient, as she had in the past, the hospital could not permit her to leave without treatment. Any treatment, no matter how compromised, that kept those babies in utero was better than no treatment that was sure to bring them into NICU by way of the ER.

After 12 days, AB refused any further treatment to delay labor. She was delivered vaginally of twin girls both under 550 grams. The babies went to NICU for an undetermined length of stay. AB went home.

This case became the base of a 20 page paper on the various "rights" of each participant. Autonomy, beneficence, paternalism – the raw material for ethics.

In Ethics class I presented another Grady experience. *Placenta Lady* is a story best told in person with hand gestures and knocks. I'll do what I can with a written description.

RB was a black female, age 23, rather large of bulk and at Grady for her first baby. She presented to the ER in the early stages of labor. She had no prenatal care or training which was not unusual.

When I came into the room she was in bed howling in pain and fear, and attended by her assigned RN, a nurse midwife and an RN midwifery student. The large room also had a couch and chair along the far wall. RB's boy friend was asleep on the couch, her sister asleep in the chair. They would remain asleep through this entire story.

RB was dressed only in her fetal monitor belt which made her look somewhat like a sumo wrestler. She was also fully dilated. With each push you could see the baby's dark black hair. The midwife had been trying for an hour to move the baby down. She pointed to me. "You look strong. Come over here and help her squat on the bed."

Squat?

RB's arm shook as she latched on to my shoulder with her sweaty fingers. The nurse positioned her squatting on the bed and proceeded to have her bounce up and down which produced a torrent of grunts, screams, sweat and palsied shaking. But no baby.

We eased RB back down on the bed. The nurses conferred. "The baby's heart beat is showing too much early deceleration."

"He could be in danger," intoned the midwife student.

"Yeah, time for the interns. We did the best we could."

First came the epidural team. The two young men in white rolled RB on to her side and stabbed her back three times (missing twice) before they correctly placed the line. RB screamed with each failed placement; they were unmoved by her pain except to note it as an indication that they had not correctly placed the line.

Then minutes later the attending physician arrived with an entourage including a surgical resident, two neonatal residents, an anesthesiology resident , and two medical students, whose function I couldn't place.

The attending was a large, gruff man who could have changed into dirty overalls and been mistaken for my auto mechanic. He conferred briefly with the two midwives who then exited stage left and never reappeared.

The two male neo-natal residents dressed in rumpled lab jackets set up an equipment station with bored laziness and never even looked at RB.

First, the attending tried to pull the baby out with suctions. He yelled at RB to push while his young lady surgical resident applied a suction cup to the baby's head. She pulled to no effect, then the attending pushed her away and he pulled. RB screamed. The baby did not move down any further.

"Hand me the forceps," growled the attending physician. One of the medical students, a young lady with uncombed hair, complied. When RB saw the stainless steel forceps, she started wailing. "Save me, Jesus!" Everyone ignored her. The boyfriend and sister slept on undisturbed.

There was a knock on the door. Who would knock? Everyone else just plowed in and out as if this were a dollar store. I was near the door, so I opened it. Facing me was a pretty young woman dressed in standard issue lab jacket, neatly starched and ironed, toting a toaster-sized ice chest in one hand.

"Hi. I'm the placenta lady. I heard you're having a birth here. I need the placenta for our perfusion research."

"Perfusion research?"

"Yes. Viral perfusion. You're the physician?"

"No, I'm a nursing student."

"Oh," she said in a drooping tone. She was probably thinking: Why am I wasting my time talking to a worm?

"The birth isn't going well. Do you want to come back later?"

"Who's that?" growled the attending.

"She says she's the placenta lady."

"That bitch? Tell her to fuck off."

She didn't blink an eye. "I can't wait long. Let me give you my card with my number. Please call me when the baby arrives. I can be over here in less than five minutes." She touched my arm, smiled and left. I put her card in my shirt pocket.

Back in the room, things, in fact, were not progressing well. The attempt to extract the baby with forceps was as unsuccessful as the suction. The neonatal residents had set up their portable incubator. One fellow stretched and yawned. They both looked as if they could use an hour or two in the incubator themselves.

The attending and the surgical resident conferred. "No choice," he said. "C-section. Let's get her into the OR."

That discussion set RB to weeping and wailing. "Don't cut me! You promised me no cutting! No scars! I can't have no scars! Please, Jesus, don't let them cut me!"

"Consent form!" the attending shouted to the RN. She fumbled through her stack of papers and found one.

He said to RB:

"TheprocedureweneedtoperformrequiresthatIinform…" The attending spoke so fast from his memorized script that I could barely understand him. "Youofpossiblenegatveconsequences…"

"Don't cut me! I need my mama!"

"…hemmoragedamagetootherorgans…"

"I want to talk to my mama! Don't give me no scars!"

"…evendeathSignhere."

"I ain't signing nothing till I talk to my mama!"

"Where's this girl's mother?" growled the attending. "Let's get this show on the road. And who are those people sleeping over there?"

"Her mother isn't here," answered the RN. "Maybe we can get her on the phone."

"Where the Sam Hell is the phone? We haven't got all goddamned day here."

I should stop here and explain about phones in Grady. There are no phones in patient rooms. They grow four feet and run away. At least that's the charitable explanation. Each floor has several cordless phones that are as hard to find as a virgin dancer in the Kit Kat Club. Being the student nurse, I knew that my present task in life was to locate the cordless phone.

Up and down the halls I walked, in and out of every room, questioning anyone with a uniform or a badge about a phone. Ten minutes later I found the phone in the hands of another huge patient who had not the slightest intention of releasing it. She was engaged in a conversation and had the phone cradled between her ear, her massive chin and her shoulder.

"Ma'm, I hate to bother you but we have an emergency down the hall…"

She rolled away from me on her bed and continued talking without dropping a syllable.

"Ma'm this is really important…"

"Gladys!" screamed a voice behind me. The RN from our room. "Time to give up the phone. I'll return it to you later with double privilege."

We returned to RB's room to find ourselves like lone passengers trying to get on a New York subway as a mob pours out from the door. We were pushed back by the surge of residents and specialists rolling their equipment from RB's room to the OR suite. Everyone was polite, vacant and bored. Everyone except RB.

The room was nearly empty again. The attending, the RN, me and RB. Plus the sleeping boy friend and sister. RB wailed: "You got the phone? Please, Jesus, give it to me so I can call my mama before they cuts me."

The attending grabbed RB's hand and guided her to scrawl her name at the bottom of the consent form. "No time to screw around," he said. "Nurse, get that bed down to the OR." Then he departed with the rest of his crew.

While the RN arranged the bed for transport, RB punched in her mother's number. No answer.

Meanwhile, there was a knock at the door. Once again I played Fishbait and answered the call. "Hi, I'm the cord man," said a tall, thin fellow with a broad smile, a quart-size cooler and a blood-smeared jacket. "I heard you have a birth here. I need the cord blood for research. Are you the attending physician?"

"No, I'm the student nurse."

"Really? How cool. An older dude as a student nurse."

I ignored the commentary. "There's no baby yet. And it looks like we're headed for a C-section. What do you need the cord for?"

"Blood research. I can't wait. Far too much to do. Here's my card. Call me when you have the cord. OK?" He turned and was off down the hall like Sebastian Coe running a four minute mile.

I put the card in my pocket along with the one from Placenta Lady. I felt like it was old times at a trade show booth. People, cards, hustle-bustle.

"You!" the RN shouted at me. "Help me with his bed."

We rolled the bed out the door, down the hall, and through the great swinging doors into the Operating Room. RB accompanied the wheel squeaking with her own vocalizations: "Don't cut me! Please Jesus! Where's my mama!"

Once we entered the operating room everything slowed down. We shifted RB from her hospital bed to the OR table. Her arms were splayed out and taped to bed extensions. She looked like a black Christ ready for crucifixion. A teal blue drape was hung at shoulder level so that she could not see the proceedings.

The cast of characters in the medical play had changed as well. At the head of the OR table on the front side of the drape were two anesthesiology residents. To the side of the table with their boxy set-up were the same two neonatal interns from the hospital room. On the lower side of the drape were the attending and his surgical resident, now both gowned and scrubbed with thick latex gloves. Also in the OR were an RN surgical nurse, a "counting" tech and a nursing student (me).

"Everyone ready?" shouted the attending. Without waiting for an answer he continued: "Let's get this baby out. You cut," he said to his surgical resident who was sweating profusely. "Cut here to here."

She made a tentative sweep. "Deeper. Damn, I hate first year residents. No guts. Now cauterize and retract. Get that stuff out of the way." That stuff was layers of fat. "This is the tricky part. Clean cut into the uterus." She must have hesitated for a second because he sharply ordered: "Do it!"

From the other side of the drape came a question from a groggy RB: "Did theys start yet? Did you find my mama?"

One of the anesthesia residents answered: "Yes, they've started."

"Oh, I don't feel nothin'"

Start yet? By this point the attending had climbed up on the table. His hairy forearm was plunged into RB's belly. Blood was everywhere. "Slippery baby. Can't get a hold of it. Get in here with me!" he shouted to the resident.

The lady resident got up on the table and straddled RB. She too plunged her arms into RB's sliced open belly. Both the docs were sloshing around inside RB like they were pounding cole slaw in a crock.

There was a knock on the door. I walked over and opened it. Actually, I needed the break from the blood and tension. My legs were turning to jelly and I was light-headed. There facing me was the Placenta Lady.

"I was wondering if you have..."

"Who's that!" thundered the attending.

"It's the research lady who wants the placenta after you..."

"That bitch! Tell her to get the hell out of here!"

"I'll just wait outside," she said quietly. "You'll let me know, won't you?"

I smiled and closed the door.

Back in the Chamber of Horrors the attending pulled a bloody rag doll from RB's belly. He handed it off to the surgical resident who was as covered in blood as he was. She got off the table and passed the limp figure with cord and placenta attached to the waiting neonatal residents who suddenly sprang to life for the first time that morning. But I know a dead baby when I see one. The morning's work wasted. Personnel, care, money, resources, mental and physical anguish wasted.

"Sew up her uterus," said the attending with weariness in his voice. How could they see what to sew when everything was a bloody, pulpy mess? I was hoping they would sew everything wrong and she wouldn't have the opportunity to kill another baby with her uncaring ineptitude.

I heard muffled shouts of anger in the hall. I pushed open the solid wood and stainless steel door to find Placenta Lady and Cord Man in a furious fight. They were screaming at each other and swinging their coolers in an attempt to smash each other.

"What's going on out there!" shouted the attending.

"It's the Placenta Lady and the Cord Man going at each other. It looks like she's winning," I answered as she landed a full body blow.

"I don't believe this shit." His large bulk moved surprisingly fast as he circled the table and pushed me out of the way. "Now you two bastards listen up. I'm sending the placenta and the cord to Pathology. You both get nothing. Now get the hell out of my operating room area." He closed the door muttering to himself. "Bastards. Give me an Apgar score," he said to the neonatal residents.

"Three."

The baby's alive? I looked more closely at the limp gray shape and saw a slight movement to the arms. The interns were still ventilating the baby with a tiny bag.

Three. A score that means the baby has severe problems – mental and physical disabilities in his future. Perhaps unable to survive past an hour.

"Sponge count," said the surgical resident. She was ready for final closure.

The bloody gauze rags had been laid out on a plastic cloth on the floor. "Twenty-one," said the counting tech.

"Wrong!" shouted the circulating RN. "We used twenty-two. Recount."

The RN and the tech count together. Using stainless steel thongs they flip and turn the long gauzes like fish on a fryer. They find two stuck together.

"You count," the RN said to me.

I did a slow count. "Twenty-two."

"I agree," said the RN. "Final count twenty-two."

"Great," answered the attending. "Close her up. I'm ready for lunch."

RB started snoring.

The next morning I visited the NICU and found the baby. I tapped the side of the crib for a startle reflex. No response. Then I touched his cheek for a Moro reflex. No response. In his chart I found that the nurses had graded him an Apgar 6 on their morning assessment. An improvement from the day before, but still not a good prognosis.

NURSING 413. Senior Year, Community Health I.

At the end of the junior year, half an hour before the final, Professor MB found me sitting in the room studying. "Henry, can you give me a minute?" she asked with her cocked smile.

We stepped out in the hall. My mind raced through everything wrong I had done in the past month. Now what did I do!

"Some of us were talking. We noticed that you hadn't signed up for the senior honors program. I was curious about why not. You're such a natural for that."

How do you tell a professor who is about to give you a final exam that you don't want the extra hassle in your last year of school. That you want to cruise through the classes and walk away with your degree. That you're tired of the academic bull shit and silly group presentations.

"My concern, professor, is that with school and work, I wouldn't be able to perform up to your expectations if I took on the added burden of the honors program." Blah, blah. My own version of bull shit.

"It's not a burden, Henry. The honor students work together as a team through the year. You would make a great addition to the program. Please think it over. Here's my home phone number. Call me tonight and let me know what you think. We could use your humor and directness. And do well on the final today."

Right.

In Nursing 310 it was as if the professor brought us to Lake Lanier, gave us a regulation fishing pole with approved 10# test line, issued us special lures, rented the boat, purchased our license, told us where to fish, and gave explicit directions on how and when to report our catch. This year professor said: "Find a pond and fish." Welcome to senior year in the honors program.

In the first quarter, my group was assigned to the Floyd County Health Department (FCHD) in Rome GA. We were here to study families rather than just the problems of individual patients. Professor K again as our instructor/facilitator. Surprisingly, she was open to most any suggestion.

"You're all seniors. This is the honors program. I know each of you. I expect you to be conscientious and thorough." Most of the group chose to work with prenatal, cardiac and COPD cases. I decided to find and work with an AIDS case and family. This may be rural Georgia, but *et ego in Acadia.*

Professor K had reservations about my plan: "I've never heard of an AIDS program here. There's no mention of it in any handouts. While I think there's a nurse assigned to STDs, in my three years here, no student has worked with her. Here's a challenge for you, Henry. Go to it."

The STD clinic was easy to find. Then I was bounced between three nurses before I found my quarry: a small office in the back, and a nurse with the singular responsibility for the AIDS and HIV cases in Floyd County.

Betty was the only black nurse I had met in FCHD. K was surprised when I introduced them: this was a Health Department nurse she had never met. Was there a link in this rural county between a county health program not mentioned in any handouts and a black nurse unknown to K and not mentioned during our FCHD orientation? Does it rain in Georgia?

Betty was pleased to work with me. "I have a family that might be perfect for you. You need to know that they are going to demand strict privacy. This isn't Atlanta. You can't tell anyone here about Jon's illness. You have to agree to that. And you have to convince them of your total discretion."

I agreed. The next day Betty arranged a meeting between herself, Jon, his mother, and me in her office. The mother was a short woman, squat of build with pepper hair and heavy, worried eyes. Jon was as pale as his tee shirt, his voice cottony, mouth darkly infected, eyes unfocused, head missing patches of hair, ulcerative sores on his arms and legs. He walked unsteadily with a cane. Did he think people would mistake him for a case of mono?

After we introduced ourselves, I explained that I was a student with experience in AIDS care and I wanted to work with them as a family unit for the next two months. "Your nurse tells me that you're both concerned about privacy. Let me assure you I will never use your real names in any report that I write. In fact, the only one who sees my writing is the professor, and she doesn't even see your real names. As a nursing student I must keep strict confidentiality."

Years ago in my New York City days, I had a friend named John who worked for Variety magazine. Everyone knew John was gay, including all of his coworkers. Nonetheless, John insisted as a prerequisite of any friendship or relationship with him that you agree to uphold this one rule: never refer to his sexual preference in his work environment. New York City or Rome, Georgia: when you want to swim in the client's pond, you play by his rules.

Since my junior year I had been a volunteer at Haven House AIDS Hospice in downtown Atlanta. I usually spent one day a week there helping the CNA or RN. This had given me a familiarity with AIDS drugs and interventions plus a closer look at the denial syndromes that are part of this and many other chronic illnesses. One of the reasons I chose to work with an AIDS patient in FCHD was my comfort level with this type of patient and my growing conviction that I would like to be an RN in this field.

The following week I visited Jon's home to meet him and his parents in the setting most real to them and most germane to our course: their home. The focus of Nursing 413 is Family/Community Nursing.

They live in a modest home, a single story frame house with three small bedrooms, two baths, and two central rooms. Maybe 1000 square feet. The yard was filled with small statuary including frogs, squirrels, and even a politically incorrect black Sambo. Hanging baskets of Boston ferns and fuscia decorated the front porch. The plain message was: someone loves to garden and grow lives in this house. Brown and red leaves blew and scattered across the street and lawn.

Inside the house was clean, tidy and cluttered with stuffed furniture and dark, heavy wood cabinets and shelves. There was hardly room to walk through from one room to the next. Knick-knacks, photos, bible verses, praying hands, potted plants, and magazines covered every surface.

Two months earlier, during the night while walking to the bathroom for his persistent nighttime diarrhea, Jon lost his balance in the front living room, grabbed for a chair, and fell bruising his left cheek. The chair he held on to was a swivel rocker that turned around as he pushed against it. On my first visit to the house, the bruise on the cheek and the chair were both still there. The family claimed there was no place else to put the chair. Which was correct in its own way, but showed a desperate lack of priorities.

Jon lives with his mother and father who are both on disability. Millie and Bill are 59 years old and have been married for 35 years. They were born and raised in Floyd County, attended school together, and have rarely traveled outside Georgia. "We went to the Grand Ole Opry onect." Millie has severe osteoarthritis that has nearly crippled her hands. Bill has three fused lumbar vertebrae from a work accident ten years ago and a history of stomach and intestinal cancer. He still serves as a deacon in a local Baptist church.

Jon has an older brother and sister who both live locally. He also had another older sister who was murdered three years ago by her estranged husband. The six year old child from that union now lives with the paternal grandparents while the husband serves a life sentence in jail. My first impression talking to the family was that I was seeing a model Appalachian social structure.

In any discussion, whenever Bill spoke first (as he often did), his wife and son listened and did not interrupt. Conversely, whenever Bill interrupted his wife or son in conversation, they immediately ceased talking and waited for him to finish before they continued with their line of thought. This pattern transported me back to my own home upbringing where my father was a strong patriarchal figure who demanded and received similar deference.

"We believe that marriage is a sacred institution," Millie said to me when I broached the subject of the deceased daughter.

"The Bible says," interrupted Bill, "that marriage obligations are for life. Not for a year or two. Forever. That's what it says. You can look it up. It's there. Now I know her husband was beating on her and that's wrong too. You have to respect your mate. But when she came to us and asked us to take her in with her little granddaughter, we just couldn't go against the Book."

Millie covered her eyes are started crying softly. Jon remained impassive as his father continued the story in a bland voice. "So we had to send her back. You got to do what the Bible says. The bond between a husband and wife is absolute. God must have had his reasons. We can't know why God does things. We don't know why He let him shoot her to death."

A nurse is not supposed to interject his judgments into the lives of his patients. But I was sorely tempted to ask Bill if he really believed that hiding behind his Bible assuaged his guilt, if he really thought that his God wanted mindless obedience over compassion and understanding. Was the truculent God of Exodus more important to him than the gentle Jesus?

Working in a hospital or a clinic, a nurse sees only a small part of a patient's life. A sanitized body in a high tech room, a family in constrained and monitored circumstances, a situation with defined patient outcomes and medical interventions. In OB/GYN how much of AB's anguish about her unwanted babies did I or the other nurses really know? Was her definition of a successful outcome better for her and the family around her than Grady Hospital's charted goals? How could I judge this from my limited perspective? How could I ever know this without entering her daily environment? Which I neither wanted nor was invited to. And in Jon's family in rural Georgia, who am I to judge their world and its rules?

I came to realize that Jon alone was not my patient. This was a whole family in chronic, debilitative physiological and mental distress. Their most dramatic problem centered around the dilemma that as the demands of Jon's at-home treatment were increasing in complexity (from pills to IVs), the parents' physical abilities to manage the interventions were decreasing. In my eight weeks with them, Bill had jabbed himself twice with a needle trying to load saline and heparin IV flushes.

Jon was first diagnosed HIV positive in 1991 after several bouts with oral candidas. He presented to the FCHD in 1992 with

AIDS. When I asked him in private about his risk factors he was adamant. "I know what everyone thinks. I have no homosexual history. None. And I never did drugs."

Right. And pigs can fly.

"Do you have any idea how you came in contact with the virus?" I asked quietly.

"Yeah. When I had my kidney removed in 1982. They gave me lots of blood. That's before they started testing blood for HIV. That's when I got the virus. And it took nine years to show up."

"Have you made any plans for your future?"

"I know, I truly know, that God has a plan for me. This is a test that I'm going to pass. He's going to cure me so that I can continue to do His Will on Earth. I believe this."

My technical mind cranked out the word: denial. But I also knew that this family, like so many others, has a pervasive religious orientation. How little time we spend on religious belief in school and hospital, and how much we encounter in homes and personal settings.

This family's home was filled with their Christianity: radio programs, clothing with mottos, statues, bibles, and inspirational books. Bill talked with pride about his work as a "lay preacher" when his ambulation was better. In Jon's room are photos of his volunteer work (before his sores appeared) with church youth.

Jon's parents are painfully aware that openly acknowledging their son has a "homosexual disease" would lead to social ostracism for them and the whole family. "One day when I was staying with Jon in the hospital," Millie said to me, "I went out for a few hours. You need a break now and then from seeing your son so sick and in pain. When I got back, the nurse asked me if I was the one who had called her asking how Jon was doing. It wasn't me. The nurse told me that's what she believed and she told the caller nothing. I guess it goes to show you that nosey people are starting to ask questions."

Jon's home care was being provided partly by a home health care agency and partly by his parents. I wasn't happy with either. I often saw dried blood on his IV site, improperly disposed bloody bandages, uncapped IV lines, a crimped access line, and dangerous injection techniques from parents lacking sufficient manual dexterity.

The family told me that a home health care nurse had said to them that Jon's walk was unsteady because his brain had already "rotted" from a CMV infection. They seemed to have little education on the medicines Jon was receiving PO and IV: Gancyclovir, diflucan, elavil, dapsone, videx (DDI), and AZT.

I spent eight weeks with Jon, Millie and Bill. I came to like them as people and respect their fortitude and steadfast commitments. I hope I lightened their burdens.

Part of being a nurse is knowing when to let patients go. You do as much as you can. Then you say a simple good-bye and leave their lives.

Nursing 414. Community II.

A course of community nursing in organizations. Yawn. Boring. Ten weeks of work in MUST Ministries with four of us passing out cans at a food bank. This is the Honors Program?

For our honors project we developed a simple way for the ministries to track their demographics. Colored pins in a large wall map. Of course, since a sizable percentage of the homeless people service lied about their demographics, the results were of limited usefulness.

Nursing 415. ICU and Rehab.

Winter quarter 1995. This hospital based course and the food bank hohum were taken concurrently. The pace of nursing classes had increased considerably. Fortunately, the hotel business that brought in the dollars to support my venture through nursing school was chugging along. No problem sewing straps and filling door stops for hotels in the hours between courses.

Many of my classmates had taken aide jobs at local hospitals. I couldn't figure out why I should make $5 per hour emptying bed pans when I could make $15 per hour sewing straps in my own home-based business. My continuing volunteer work at Haven House had left me with considerable confidence that as an older male RN, I would have minimal trouble finding a job. The market for RNs was shrinking with the growth of managed care around Atlanta, but the pool of male RNs wasn't growing very quickly.

I hated Rehab at my assigned hospital. To begin with, this is a hospital I would never place at the top of my list for patient care and nursing competence.

As part of the requirements for the course, we had to write a paper on a "Paradigm Experience". My paper was based on a small incident that is outlined below.

Mr. N was a 68 year old CVA (Cardiovascular accident) patient with right side weakness and occasional confusion. He had been on the Rehab Unit for nearly a week and was not scheduled for discharge for another week. Mr. N has adult onset diabetes for which he takes insulin on a sliding scale before every meal.

Just before lunch was served in the dining room (most patients do not eat in their rooms on Rehab), my nurse supervisor asked me to assess Mr. N's glucose level by a hand-held blood glucose monitor and administer regular insulin if needed. At the nurses' station, I accessed Mr. N's chart, checked the doctor's typed orders, and checked the nurse's notes to determine that Mr. N had his pre-breakfast insulin injection subcutaneous in the right abdomen. I performed and entered a quality control (QC) run on the hand-held monitor, and went to the patient's room.

Mr. N was not in his room. Instead, I found him seated in a wheel chair at a table in the dining room waiting for his lunch. Who had assisted him out of bed, into the wheelchair and into the dining room?

"You need to check your glucose level before you eat lunch," I reminded him.

"Oh, that's right," he said quietly.

"Do you want me to check it here or in your room?"

"Oh, not here. In my room, please."

I wheeled Mr. N back to his room where I drew blood from his right middle finger. Glusoce of 256 on the machine. Far too high for him to be eating lunch without insulin.

After making sure that Mr. N was strapped into his chair, I returned to the nurses' station to double-check the orders. The scale indicated that for a blood glucose of 256, he needed 3 units of regular insulin. I drew up the units in a small syringe, and brought the needle to my nurse for her review and approval.

Back in the room, I gave Mr. N the 3 units under the skin (SC) in the left abdomen. Routine nursing.

"Are we done yet?" he asked politely. "I'm hungry and ready for lunch." Then the problems began.

"Almost," I answered. "Let me get rid of this needle. Then we can go." I looked and could not find a "Sharps" box anywhere in the room to deposit my contaminated needle. Or in the bathroom. I poked my head out of the room and saw a nurse's aide I recognized from the day before.

"Millie, I can't find a Sharps box in this room."

"There aren't any in the rooms," she said. "You'll find a red box on the side of the med cart."

"Where's the cart now?" I asked, the needle held upright in my hand.

"It's either in the room behind the nurses' station, or it's down in the dining room." Places at opposite ends of the hall.

The situation left me with a set of options, each of which

broke a nursing rule. I'm not supposed to cap a used needle: it is strictly against universal precautions, it places me in too much danger of sticking myself. Nor am I supposed to walk around the halls with an uncapped needle: too much danger to others. Nor am I supposed to leave an uncapped needle on a patient's side table while I wheel him off someplace.

In the end I opted to cap the needle and assume the danger. Then I had to decide between four plans: leave Mr. N in his room while I found the Sharps box and got rid of the needle, or leave the capped needle in the room while I took Mr. N back to his lunch, or take the capped needle with me in my hand or in my pocket while I took him back to the lunch room, or press the call button and ask someone to watch Mr. N as I went to locate the Sharps box.

I told Mr. N that I was going to find a Sharps box, and I would be right back. He shook his head. "Tsk, tsk," he said. "My regular nurse never has these problems."

The med cart was behind the nurses' station. I deposited the capped needle in the red box, and returned to the room to find my patient back in a pleasant mood. We sauntered down to the lunch room, me pushing and him free-wheeling.

Later in the day when I explained my plight to my nurse, she said: "We don't have those boxes in the rooms here in Rehab. Maybe it's not right. But we don't give enough injections for a box in the room. At least that's what they tell us."

The incident sounds pretty trivial. And as a floor nurse I would have probably not been bothered much. However, as a student trying to do exactly the right thing by the book, it was a trial of compromises I was not comfortable with.

Nursing 416. Preceptorship. Spring 1995.

What a long road. Final quarter. Final course. Preceptorship
is a one on one skilled nursing with an experienced RN. Eight
weeks of at least 24 hours per week. Most students work at a
hospital in Med-Surg. I asked permission to precept at Haven
House AIDS Hospice.

Professor R was my clinical advisor again. Not that I minded
since after two courses we knew each other pretty well. Two
seniors (Kathy and I) from Kennesaw rotated preceptorship at
Haven House (I set up the program). Sometimes we worked
individually, sometimes as pairs. As part of the course, we had to
keep a diary of our actions and thoughts. Below are excerpts from
that diary.

April 14, 1995, Tuesday.

Vara (the RN) assigned Kathy and me five patients each. I
have the back of the house. There are five rooms in this old house,
two patients to a room. No effort was made or intended to make it
look anything other than a house.

Vara is tall and thin with long, kinky black hair. She could be
very attractive if she didn't dress in such sloppy clothes or if she
put some animation into her aspect. On the other hand, she is one

of the most competent nurses I have encountered.

Today started at 7 am with Mike complaining of nausea. Mike is a 30 year old six foot tall skeleton covered with skin, indistinguishable from a prisoner at Auschwitz. TPN (total nutrition by IV) at 91 cc per hour continuous had not stemmed the drop in body mass. His nausea is best treated with a 25 mg Phenegran recal suppository. Not enough muscle for an IM injection. Vara looks at me and says: "Go to it." I hope this first intervention is not an omen of what the next eight weeks will be like.

Mike has two stage I decubitus sores, each the size of a fist. One over the knobby prominence that was once called his left hip and one on his equally defined sacrum. He came from the hospital in this condition.

The staff strives throughout the day and night to keep him turned on his right side, propped with half a dozen pillows, to relieve the pressure on his sores. He fights us by constantly turning back to his left. His eyes and his mind are clear. He knows that he is causing damage.

That afternoon, as on most afternoons, his parents come to visit. We move him out of bed into a chair. He motions and grunts for his cigarettes. The three of them sit in the main room in separate chairs. Mike smokes a great cloud. I don't hear a single word pass between them.

Every afternoon between 3 and 3:30 we have a change of shift conference. Today Vara, Kathy, Cliff (the evening RN) and I meet outside at a round, rusted iron table. During our talk, Mike's father walks by, heading for his car. We wish him a good evening and he smiles back. A few minutes later, Mike's mother passes us, her shoulders hunched together, her purse clutched to her chest. We say good evening to her. She looks up with a tortured face and merely nods. I felt her anger; it was hot like my pot belly stove when I lived on the farm in New York State mid-January.

I did not see the parents sit closer than half a room away, did not see them touch or exchange a word. The father occasionally conveyed Mike's wishes to us: move the chair, get a pain med. Vara said no staff member has been successful penetrating the mother's wall of anger and denial.

In a hospice, the resident is not your only patient. A nurse must extend his care to include the family, significant others, and close friends who visit smiling and upbeat and who drop their faces and bravery in the hall. In the morning you practice med-surg with the patients. In the afternoons, when friends and family visit, you practice community and psych nursing. That's the territory.

Just before we headed for the camels and disappeared into the desert, Vara and I returned Mike to bed and changed the duoderm over the hip decubitus. The flesh red, inflamed and nearly worn away. The bone was directly underneath with little tissue coverage. Vara spoke sternly to Mike. "Listen to me, Mike. Unless you make an effort to stay on your right side, a real effort day and night, unless you do this, the minor discomfort that you feel now with a stage one decubitus will be nothing compared to the pain and infections that will come when the skin breaks down and the sore infiltrates."

Vara is very stern with Mike. He's depressed and angry, not stupid and confused. Mike flashes her a glance and looks away.

Nathan is one of my favorites. I've known him for almost a year. When I first met Nathan he was a medium build 40 year old black male suffering from mild confusion (AIDS dementia). Blessed with the gift of the fine phrase, sometimes cute, sometimes raunchy, he had his way with the nursing assistants (CNAs). Unfortunately, his decline has been slow and inexorable. He was now mostly confined to bed, blind from cytomegalovirus (CMV) infection, hardly eating, and only recognizing us in flashes.

Nathan has a portacath mainline access located implanted under his skin in his right chest. You access this port with an external needle (a Huber needle) that passes through the skin and stays in place for three days at a time. It becomes a regular nursing chore to remove the old needle, clean the site and insert a new needle.

One morning Vara told me to perform the sterile procedure while she watched. I spread out my sterile field on Nathan's belly and spoke to him. He stared vacantly.

The intervention proceeded uneventfully as I pulled out the old needle, swabbed the site with alcohol and Betadyne, and prepared the new Huber needle for insertion. Nathan screamed as I touch the needle to his skin. I withdrew my hand.

"Try it again," said Vara.

I barely touched Nathan and he screamed again. We both know that he can't really feel anything yet. He was like a dental patient screaming when the drill turns on.

"Just do it. Push it in," she said with added emphasis. Nathan screamed and flailed his arms in the air as I started again.

I was sweating and felt faint. My legs turned to jelly. Not since a particularly messy C-section at Grady had I felt this far out of control.

"You're going to have to take over," I said as calmly as I could to Vara. She started to object and I cut her off. "No. You have to take over. I'm past my limits." I stepped back. I knew that beyond where I was lay error and mistake. I was willing to take her scorn and anger before I made a mistake.

We had to hold Nathan down as he screamed and yelled

when Vara pushed in the Huber. Later, with Vara seated on her stool at the nurses' station by the kitchen, she told me: "I never heard anyone object the way Nathan did. You barely feel the needle. We'll have you try another Huber change on Thursday with Mark. He just lies still."

I drove home convinced I made a fool of myself. I had proved that clinical nursing is not for me. I was depressed about having to quit school and finding a way for people not to think poorly of me.

April 5, 1995

My second day began the same as the first: Phenegran suppository up Mike's chute. This time I found him positioned on his right side. Had Vara's straight talk permeated his stubborn skull? I hope so for his sake. Major decubitus sores are painful for the patient and disgusting for the staff.

Right after removing my gloves from the suppository run, Keith thrashed around and fell out of bed. The back two rooms of Haven House are like a psych ward with three severely confused patients (Nathan, Keith, and Thom). They are also nearly blind from CMV. Fortunately they're in the two rooms nearest the nurses' station so we can keep an eye on them. Even so, wandering and falling have become so serious that today we put them in Posey chest restraints.

Restraining patients is not the preferred procedure and goes against the spirit of a hospice. What can you do when the patient is so out of control that he becomes a danger to himself and others? Vara suggested putting their mattresses on the floor. That would eliminate the writhing around and falling out of bed. But how do you prevent the wandering? And falls in the bathroom and hallway?

This morning I help Linda and David, the CNAs, bathe and dress my five guys. This way I can give them a full body assessment on the physical side and judge their mental states as we talk to them and ask them to move or help us. Linda is a huge black lady with an endless compassion and a motor mouth to match. David is small, sardonic, very gay, and often contemptuous of his position and the patients. He's good at his tasks but he lacks Linda's commitment.

Thom is (was) a pharmacist. When lucid, he's a delightful person with a quick wit. After we clean him up and I make notes on his physical state, we put him in a chair in the sitting room with the morning news, some other fellows and the general activity of the house. Because he's so unsteady and often confused, we restrain him with a vest to the chair. The nurses and Thom are aware that the next step past the restraint is medication and sedation.

What will we do with Thom when his blindness and periods of mental confusion become more intense? Do Haldol and Mellaril enhance the quality of life if they prevent falls? Is a fog better than acute distress? Where is the line between adequate safety/care and loss of identity?

At Haven House the nurses have to face these ethical questions and make judgments. You can't turn to a doctor and say: "Tell me what to do. Give me a written order." We have the standing orders for multiple meds. The RN makes the decision. It goes with the territory.

At this afternoon's conference, Vara, Cliff, Kathy and I along with Mona, the over-all nursing supervisor, discuss these issues. Vara argues for greater restraints citing Nathan's latest fall that resulted in a large cut over his eye. Cliff proposes greater nighttime sedation citing Thom's violent episodes if his Haldol levels fall too low. Mona, who could pass for an underweight ten year old boy if she removed the multiple ear rings, leans toward greater restraints with increased supervision and bedside time. Never during the discussion does anyone talk about law suits or cost/benefit ratios. The talk and the decisions are about patient care and quality of life.

April 6, 1995. Thursday.

Kathy and I were the students this morning. Vara asked me to take care of the four men at the front of the house today: Tony, Mark, Jim and Mike. Mike was moved from a back room to the front room with the wide views. Just in time for me to care for him again.

Kathy has the three demented warriors in the back two rooms: Thom, Nathan and Keith. She also has a new patient, Horton, who was admitted yesterday afternoon on a transfer from Crawford Long Hospital with a blood oxygen saturation (O2 Sat) of 81 (Normal 98), febrile, panicked and confused. What has happened to our hospitals that a patient could be released in such unstable condition?

When I first looked in on Mike my nemesis skeleton, I saw that he was curled up on his right side (yes!). However, when I checked closer, I found his eyes were glassy and his talk rambling. He did manage to ask me for some Pepto Bismol for heartburn and to complain of nausea. My regular way to start the morning. Henry, Suppository Man. I inserted the Phenegran and gave him 10 ml of the pink elixir. Then I stood by the bed and waited for the inevitable.

As expected, he burped and spit up the whole mess. Because he's on TPN plus lipids, he only spits up the Pepto Bismol plus some thick mucous. I turned his skin-over-bone skull to the side to prevent aspiration, wiped his face and sparse hair with a cool cloth, and reported my success (sic) to Vara.

She's sitting on a stool between the nurses' station and the kitchen, eating a bagel with cream cheese, and drinking from a mug of coffee. Vara loves being a preceptor. "Hold his morning meds. Ask David to call us when he's ready to clean up Mike so that we can give him a full evaluation."

Later in the morning, I take out a few minutes to talk to the new resident, Horton. His color is better; on admission last evening his color was almost blue. He is alert and more oriented. He has active PCP (Polycystic pneumomnia). Even with his Oxygen set at 5L per hour, we don't expect his O2 blood saturation to rise above 85. But we can't let it fall much below 80. That becomes very risky.

"It feels great to be clean," he says to me with a wan smile.

"Clean?" I answer. "You weren't clean at Crawford Long?"

"Well, they washed my face. You folks really scrub me down. And give me fresh clothes. I haven't felt this clean in weeks. It does wonders."

Simple, basic Florence Nightingale nursing. So much for MRIs and EKGs and all the other high-tech hospital equipment and interventions. This patient values cleanliness and comfort.

April 11, 1995, Tuesday.

"Horton died on Saturday, honey," Linda said to me when I walked in this morning. "Mm hmm, he's gone." What a disappointment. He looked so much better when I left on Thursday that I was sure he would be one of our turn-around stories. Instead his O2 Sat deteriorated to the low 60s by Friday. Around 65 is the line of no return. Once below that line, a patient's death is generally inevitable.

What was not inevitable was the discomfort and gasping for air that I was told marked his last 24 hours. The weekend staff was unwilling to give him more Ativan than the Q8 hours specified in the orders. Mona, Vara and I were very disturbed to learn that efforts were not made to keep Horton as comfortable as possible.

Haven House is a skilled care hospice, a phrase that is almost an oxymoron. The staff provides intensive, interventional medical and social care designed to improve the health of the patients. However, most of the patients die in-house, which is their expected outcome.

This terminal aspect of care calls for a palliative and often non-interventional approach. The staff has the duty and the trust of the physicians and families to know where the line is drawn between interventionist measures and palliative measures. When Horton crossed the line on Friday, palliative measures, not strict chart reading of med administration, should have predominated.

Two of my patients are approaching the line this week. Jim continues to decline physically and mentally. His first decubitus sore has appeared this week. Jim is a walking skeleton when six months ago he was a robust alert man. His physician has refused to put him on TPN and Jim will not change physicians in spite of our urgings. ("What's the point?" his office nurse said on the phone.). Jim's fears and denials have grown so large that it's hard to rationally communicate with him.

I like Jim. He's quiet spoken and thoughtful. Rarely demanding. Every morning I arrive at 6:30 am to find him already in the main room chain smoking and watching the early news. I realize that I will have difficulty when my job as a skilled care nurse changes to a palliative care giver.

Recently I read the book by Kubler-Ross "AIDS: The Ultimate Challenge." I do not approach death with her detached acceptance and her belief that death is a passage to another level of life. To me the losses of death are forever and absolute.

Vara insisted that I change Nathan's Portacath today. I succeeded this time. We went through two sterile dressing kits, two Huber needles, and three pairs of sterile gloves. Most expensive needle change in Haven House history.

Nathan screamed and yelled with unrelenting vigor through

the procedure. But I put the needle in correctly (on the second try). Later, on the porch, I shook like a ginko leaf in the wind. David saw me through the window, came out, and put an arm around my shoulder. "Hey, you done good. In a couple of weeks, you'll be a real pro. Just wait."

April 12, 1995. Wednesday.

My mother's birthday. She who also experienced a slow, degenerative at-home hospice death. I mention this to no one.

My second patient crossing the line is, of course, Mike the skeleton. Even with his TPN, he continues to lose weight (if that's possible). His sores have worsened with stage II skin breakdown beginning over both hip bones. The multiple sores over his tail bone have opened and the tissue is turning black and yellow with necrosis.

Unlike Jim, I have no personal affection for Mike. He is an angry, nasty person who deliberately coughs and sprays the nurses, refuses to cooperate with his care, and has an on-going fight with his father and mother.

He has often vocalized that the first shift does not like him. He is correct. What's there to like? I suppose some of my professors would want me to be like the nurses who write articles in "Nursing 95". Caregivers who go to extraordinary lengths to reach the positive side of their patients. Well, I don't believe Mike ever had a positive side. He certainly doesn't have one now.

I asked this question to Professor R who specializes in ethics: Can't I be a quality nurse if I give the best quality medical care and make no additional fruitless efforts to be personal to a patient who neither wants nor accepts them? How much of my energy and patience can I expend with one patient when I have others equally needy?

R said to me: "Beneficence versus justice. There's never a tidy answer. Do you know why he's so angry?" No, I don't and I'm not sure I care. Or that it matters.

I don't neglect Mike. Yesterday and today I rolled up my sleeves and helped David and Linda with Mike's bath and morning care. Mike rewarded me by blowing smoke in my face when I was reattaching his TPN line after we had moved him to the main sitting room.

Mark on the other hand is a quiet man in relatively good health. At first glance (unlike Jon from last semester), you would not know he is quite ill. Only his clonky gait gives away his CMV brain damage. Mentally he is as alert, bright and oriented as I am. Mark does his own morning care, sleeps late, comes and goes as he pleases, and depends on us mostly to regulate and administer his medications. Mark also has a Portacath for his IV Gancyclovir.

Vara added Urokinase to Mark's plugged Portacath today using his IV line as the access. She let Kathy and I observe the procedure but not participate. "Maybe later in the semester, if we have another." An hour later she removed the clot buster in a reverse procedure by drawing up the med plus considerable blood into a 10 ml syringe attached to the end of the short IV tube exiting the Huber needle.

She performed both procedures without gloves. Twice I passed her a spare pair of latex gloves from my pocket stash; twice she refused. Vara's procedural skills and knowledge are remarkable. But her safety skills, at times, are deplorable. Chicken Henry did not confront her. Maybe later in the quarter.

April 13, Thursday.

Mark developed a fever last night. Today we had to repeat the blood draw using both ports of his dual sub-skin portacath. The concern is that if the infection is in only one of the two ports, we would have missed it yesterday with the single port draw. Vara let me perform the draw on one port and gave Kathy the other. I used gloves.

Mike the Skeleton was better today on the social side but worse physically. Yesterday afternoon Vara had another strong talk with him, this time about mistreating staff. Then she growled at me. "Henry, let me tell you this: never patronize a patient. When their behavior is inappropriate, treat them like adults and tell them like it is."

Mike and David (the CNA) had a rough day yesterday: Mike was obnoxious, David unsympathetic. So this morning I put on my CNA hat and did his basic morning care. I also gave Mike some extra care: lip balm, double oral swabs. Maybe the time and concern did some good because he was cooperative with me during his Huber needle change. Which, by the way, went quite smoothly and earned me some much needed praise from Vara.

On the down side, we couldn't convince Mike to leave his bed and sit in the family room. That's the only place he's allowed to smoke, so staying in bed carries a weighty price tag. This is the first day Mike has refused (unable?) to leave his bed. Not a good sign.

Mark came to Vara and me late in the day and asked why Jim was not on TPN when he obviously needed it. Vara and I looked at each other in surprise. This is the first time Mark has shown any concern for anyone other than himself.

"Jim isn't on TPN because his doctor refuses to write the orders," I explained.

"Why doesn't he go to my doctor? She'll write the orders."

"We've asked him to change. He won't do it."

"Maybe I'll talk to him," said Mark.

"Great," we both responded with enthusiasm. We like Jim. His slow wastage is troubling medically and painful on a personal level.

Roger is one of my favorites. He's a Haven House success story. Very ill six months ago, he's nearly self-sufficient today. Like Mark, he does his own care, keeps his own schedule. We monitor and administer his extensive list of medications.

Several months ago when I was cleaning his central lines where they exit his chest just above his right nipple, Roger said to me: "Oh, Henry, You rub my chest so good. I could become passionate toward you."

"Sorry, Roger. I'm happily married."

"Oh, such a loss."

Today, Roger pulled me over in the hall and commented on Mark. "I am the queen of Haven House. The true queen. Not the others who think so. You know who I mean." Great. Now we have a battle for the queen's crown.

April 20, 1995.

Once again today I change into my alter-ego: Suppository Man. I'm probably the only nurse in Atlanta who can identify his patients from the bottom or the top. And can tell you about their sexual preferences.

Our fellows are prone to oral problems: Candidiasis waiting to take advantage of the mucous and reduced saliva; aspergillus looking for warm, dark, moist and friendly places. One slack day and the junk accumulates. Plus, my fellows don't want you clean out their mouths even though they feel great afterwards. It's easy to slack off on oral care. And Vara correctly growled at us this morning.

Another day as intense in the psych as the medical nursing. Keith started the morning calling for help. I went to his room.

"Keith, this is Henry. What's wrong?" I touch him on the shoulder. He's nearly blind from CMV. We have had to restrain him. He kept pulling out his IV line which causes a bloody mess.

"They're trying to get in the doors."

"Who?"

"The rabbits."

"Let me turn on the light and look behind the door...I don't see any rabbits here. I'll keep the light on anyway."

Keith called out again about ten minutes later. Repeat above lines. Then he said: "Can you get me some scissors?"

"Why do you need scissors?"

"I have to cut off these gloves."

"After I give you the IV Gancyclovir, we'll wash you up and

take off the mitts. But you have to promise not to pull out your line again today. Is that a deal?"

Nathan (of the infamous Huber needle change) has been cooperative this past week in spite of his growing blindness and confusion. However, this morning he became quite agitated as Linda went to feed him.

"That Nathan, he kinkin' up this morning. Won't do nothin' for me. Where's his Haldol. He got to have somethin'."

I checked the med sheet and found that he has a standing PRN order for 2-5 mg Haldol, IM. No way am I going to approach Nathan with an intra-muscular injection. Vara and I discuss the problem and decide to deliver 2 mg of Haldol diluted with 3 ml of saline through his IV line.

I prepare the med and two saline flushes. "Nathan, I'm going to give you some Haldol to calm you down."

"Yeah?" he says with suspicion, looking at the ceiling with blank eyes.

"I'm going to use your central line."

His eyes open wide in terror. Does he remember how I botched his Huber needle change last week? "You're going to hurt me! You're going to hurt me with that needle!"

Before I take his ranting personally, I remember that I dealing with a patient in dementia. So I approach him as if he were my six year old Oxy from Grady again.

"Nathan. No needles. I'm not going to hurt you. I promise."

"Why don't I believe you?"

"Nathan, listen to me. I'm holding the port at the end of your central line. I promise you no needles. I'm using a syringe with no needle on it. Let me tell you what I'm going to do. First, I use one saline flush to rinse out your line..." I describe the moves step by step. "That's the end. We're all done. I didn't use any needles. Did I hurt you? Did I keep my promise?"

"Yeah. This time."

April 25, 1995

Applesauce and pudding day at Haven House. Three of my fellows (Tom, Nathan, and Keith) can no longer easily tolerate solid foods or take down meds by mouth. For example, when you want Nathan to eat, you have to instruct him to chew and swallow, Over and over. Even then, you can't be sure that the food has been properly masticated or swallowed. Very time consuming.

So now I grind up their meds and mix them with applesauce or vanilla pudding. Of course with Tom it seems pointless to give him Gancyclovir by mouth (PO) when he's wasting away.

This morning I noticed that Nathan had a redness in his completely blind right eye. There was no drainage, and he did not complain of pain. The situation did not seem serious enough to report to Vara, but I did make a mental note to monitor the eye through the day. By noon the eye was watery, red and somewhat swollen. Nathan flinched when I lightly touched the upper lid.

I reported my findings to Vara who was munching on a sandwich. "Doesn't sound like much."

"I agree it's not top priority. But I think it should be more fully evaluated."

"Nah, he probably just poked his eye with his finger," said Vara licking dressing from her finger.

"That does sound likely. But I would still like you to evaluate his eye."

"OK. In ten minutes."

After inspection, Vara stayed with her opinion that he stuck his finger in his eye. I charted what we observed, and at 3pm report I passed on the findings with a recommendation that they watch the eye through the evening. I felt good about being a patient advocate and insisting that Vara respect my judgments even though she disagreed.

On the porch Linda told me that she has seen Nathan scratch his bottom and then move his hand to his face. Infection highway.

Today with four patients I gave meds PO, Sub Q, IV, IV Push, rectal suppository, and ophthalmic ointment. Prepared meds by crushing, diluting, spiking, and greasing.

I sat in the family room this afternoon and spoke with Roger about social problems around the house. The queen of Haven House is always available for comment. Also tried to convince Jim to change doctors; he is unwavering in spite of his growing problems with nutritional deficits.

Late in the day I sat on the porch with Mark and chewed the fat about his hopes to leave the hospice in a few weeks. He might succeed if he continues to improve. Also chatted with Linda and David about patients and spring vacations.

Mike's father came out to enjoy the glorious spring afternoon. And I fed him a pile of bologna about how Mike, in spite of his skeletal appearance, is a real fighter. Then I gently asked if the family had made any plans... He said that his wife wasn't ready yet.

At 4 pm I walked out the front door, down the wood steps and across the flagstone walk feeling that I had done a great job. Yes!

April 26, 1995. Wednesday.

Unexpectedly, most of my day was spent caring for Keith whose health took a nosedive in the space of an hour. During the early part of the day he was mumbling, agitated and confused as usual. We had him restrained with a Posey vest and mitts. He hardly knows they're on anymore, and without them he pulls out whatever tubes and lines he has. I gave him his meds crushed in vanilla pudding along with Compazine IV push after his Gancyclovir IV.

At 10 am Ken, his significant other, came by to visit. Ken told me he had been in last night, and Keith was very confused then too.

"Keith's mother and father live in Birmingham," Ken said. "They plan to come for a visit this weekend. Do they need to come sooner?" His voice was barely steady.

"I don't think so," I answered. "He's been like this all week."

Half an hour later, Linda, David and I send Ken out of the room to wait while we clean, dress and do a head-to-toe assessment on Keith. We had put Keith off for last so that I could spend extra time in assessment. Right away I noticed that Keith's feet were cold and his pedal pulses absent.

This was strange because I assess feet every morning at 7:30. I stand at the foot of the bed and chat with my fellows briefly while I check for pedal pulse intensity and balance. Nothing was wrong with Keith then. Linda said that she didn't notice anything unusual during her regular vital sign check at 7 am. The rest of Keith's body was warm to the touch. An axillary temperature showed 100.5 F.

His hands were also cold and clammy.

Keith has a Foley catheter and a diaper. David starts to change the diaper while Suppository Man readies his ammunition. I watch as he finds a small, black, tarry stool.

"That's it," David says.

"Yup," Linda agrees. "He's dying. Won't make it though the day. Better call the family."

I check his mouth. There's some blood on the gums. There's also blood on his mitts where he's been chewing.

We call in Vara and Mona. Mona wants to check the O2 SAT in his fingers. She reaches over and starts to remove the tape from the gauze mitts.

"What are you doing without gloves?" asks David. "Henry, give her some gloves right now. Mona, get your hands away from there without gloves. Don't you see the blood on that mitt?" David is not afraid of the Director of Nursing or any other positional authority that I ever noticed. Mona does as she's told. So do I.

Finally we check the O2 SAT: 72 on the fingers, 78 on the toes. Mona orders O2 at 2ml/min by nasal canula. That's the easy part.

I strip off my gloves and go outside in the hall to talk to Ken. I see Ken as an extended patient to Keith. They are both my responsibility. "It looks as though Keith has taken a turn for the worse. I think you ought to call his parents and tell them to come on over. I know this has not been easy for them or for you. But I'm a parent; and I would want to be with my child if it looks like he may not be able to pull through an episode of this disease."

Ken makes the call and then returns to stay with Keith. Keith's mitts are back on, but I remove them so that Ken can hold his hands. "It's OK to keep the restraints off as long as you're here with Keith. Don't leave him alone. Call for one of us if you need to step out. He's pulled the caps off his IV access before, and now he'll pull out his Foley and O2 as well."

Ken's crying, and he nods his head in understanding. His hands are clamped around Keith's. I put my hand on Ken's shoulder and feel his shaking sobs. "How are *you* doing?" I ask.

"I'm all right," he says. I know he's not. Why should he be?

At noon, Kathy, Vara and I go out back in the dappled sunshine to eat lunch at the picnic table. We talk meds and side effects. One bite into my noodle casserole and David is at the back door calling to me. "Ken's really thrashing around in there. Is there anything we can give him for comfort?"

Vara and I agree on Compazine IV and Benadryl PO which are standing orders for Keith. I draw up the meds and flushes. Fifteen minutes later I'm back outside.

A few minutes with my cold noodles and veggies. David

again at the back door. "We just took his temperature. 103.5 F axillary." I mumble the S word.

Ice packs for under Keith's arms and cold compresses for Ken to hold on his forehead. Keith's feet are still cold, clammy, and virtually pulseless. And it seems to me that the cold has moved up his legs past his shins. My imagination? Fortunately the finger O2 SAT is up to 90, and the digits feel warmer.

Unfortunately, my noodles, neglected and soggy, are colder than ever. I finish them alone out back under the oak trees.

Keith and Ken stay laced together as I leave that afternoon.

April 27.

Keith did not die last night. That's the problem with AIDS and young men. They're so strong that they don't die quickly the way an older patient would. Death comes like a Yukon sunset, a long twilight as the sun skims the mountains.

I spent the day as a hospice nurse giving cleanliness, comfort and dignity to a body slowly shutting down. I cared for him, his family and friends, extending my nursing role to others as his disease slowly constricts his life. Not a pleasant day, but rewarding in its own way.

May 3, 1995.

Both Tom and Keith died over the weekend. Their deaths, according to Linda were like sighs. They died lethargic and confused, surrounded by family and friends in a home-like atmosphere. A new patient, Gene, came yesterday to take over Keith's bed next to the window.

Mike the skeleton died today alert, oriented and petrified with fear. When I came in at 7, he was sweating profusely with labored, stentorian breathing like a wet locomotive, and he was groaning with eyes wide open and fully dilated. His Morphine pump was set low and his Dilaudid as needed (PRN) order was for 2 mg PO. Lot of good an oral med route does for a man who has been on IV TPN nutrition for a week because he can barely swallow. I crushed his Dilaudid in orange juice and dripped the solution into his mouth. Then Suppository Man popped a Tylenol up the chute. I noted that he wasn't a virgin in that department.

I worked with Linda and David to clean Mike. His sores were awful: necrotic, postulant, deep. We changed his wet-to-dry dressings. Although my Dilaudid drip and Tylenol suppository had reduced his pain, he moaned loudly when we turned him. David mumbled his approval at the pain. Yesterday Mike had spit at David during his morning bath.

Finger O2 SAT was 66, dangerously close to my magic no-return number of 65. Breath sounds "junky" bilateral. I suctioned his throat and put him on O2 4 ml using a face mask since he always resisted the cannula, and he was only breathing through his mouth now.

Out in the hall Linda said to me: "This man's goin' soon. Should we call the family?"

I thought about it. Every day his parents come to visit. The father sits in the corner chair. The mother sits next to her son's bed. No one touches or says a word during these visits. Two days ago Mike spat in his mother's face when she bent down to kiss him.

"No," I answered Linda. "We wait. They'll be here this afternoon. I'm not going to put his parents through any more pain than they are in already."

I presented the problem of Mike's continued discomfort to Mona. "His pupils are dilated. He's groaning in pain. I think his Morphine pump is set too low. In any case, he's end stage. We need to focus on his comfort."

"Well, that's thinking like a nurse." Acting on nebulous authority, she doubled the Morphine dosage. By 10:30 he was resting comfortably and his O2 SAT was up to 80.

At 11:00 he was once again agitated, panicked and sweating profusely. His breathing had changed to an irregular Cheynes-

Stokes, his eyes were rolled upward. However, he was alert, responsive and terrified. His Ativan and Pamelor were still listed as PO only, but I remembered several weeks ago (seems like years) that we gave a patient an Ativan pill under the tongue.

I left the room to check the med book and bring back an Ativan. I was at the nurses' station when Linda thumped down the hall and said to me in hushed tones: "Henry, you better come to Mike's room. I think he's gone."

No question. Quiet as a stone. No heart beat. Linda goes to find Vara who agrees that he's dead. We remove all his lines and tubes. Straighten him out in the bed and neatly arrange his sheets. My first patient death.

Let's have no crocodile tears here. Mike was a hateful person. "Can't say I'm sorry he's gone," says Linda. I nod my head in agreement.

David comes in and smiles. "It's about time."

"Glad we didn't call the family," I say.

Linda nods. "No reason to torture them any more than he did. That poor woman sitting there whiles he spits on her. His own mother!"

Strangely, he had some of the most pleasant visitors. They trickled in during the day as usual asking to see him when they only found an empty bed in his room. Since I was the nurse on duty (albeit a student), I took on the chore of explaining the events of the morning and comforting the weeping friends. I felt no empathy since I couldn't figure out what the weeping and sadness were for. This was a singularly dreadful person.

After lunch, Mike's father came in. His burden and sadness written bold in lines across his face. We went to the front porch. We sat in wicker chairs in the lambent sunshine. Watched the cars drive by on Fourteenth Street. A lovely Southern Spring day.

"It's odd," he said. "This morning I went to the funeral home and made arrangements. The first time I did that. His mother said: 'Anyplace but the funeral home on Alpharetta highway that we pass everyday. I couldn't bear that.'"

"Mike was conscious to the end," I said. "And it was so fast that we didn't have time to call you. He wasn't in any pain that we could see. And he was certainly the toughest fighter I have ever met."

"Yes, that he was. We're really grateful for the care you gave him."

I mumbled a few more comments, my concentration on his reactions. Mike's father was now my patient. I didn't feel any remorse for my shaved ethics. Book learning has a place. So does common sense.

May 16, 1995.

Last day of clinical for Nursing 416. Last day of clinical for nursing school. Amazing! What a journey from that first day in the hospital when Angel and I couldn't figure out what we were supposed to do with an aphasic elderly woman with mistaken dentures. Today I took over the care of all eight patients in the house and managed their interventions with ease and compassion.

After lunch I sat down in the nurses' area with Mona and asked her about the possibility of coming on board after I graduate and pass my RN boards.

"You have the energy and attitude that we want around here. You will certainly be considered. Please submit an application."

Linda told me this morning that I was the best student they had ever worked with, the only student who worked with the CNAs and evaluated every patient head to toe every day.

Now that's what I call a good way to end nursing school: Leave my clinical location feeling like I've enhanced Haven House, performed throughout the semester up to my inflated expectations, and have the facility promise me an employment consideration.

Epilog

National RN boards (NCLEX) frightened me. I studied for two hours every day for two months. It was a computerized test that could be anywhere from 75 to 265 questions, multiple choice. My test was 75 questions. I was overwhelmed by the difficulty and paralyzed by simple questions I should have known the answers to without a second thought.

Morose and dejected, I went to the college and found Professor B in his office typing on his computer.

"I blew NCLEX!"

"How's that?" he asked without looking up.

"I didn't even understand the questions."

"How many questions did they ask you?"

"75."

"Were the last two questions easy?" he said finally looking at his panic-stricken student.

"Those were the only easy questions!" I blurted out with tears in my eyes.

"Go home. You passed." He returned to his computer.

I didn't take the job at Haven House. Or at Grady AIDS Clinic. Instead, on the day my RN license arrived in the mail, I accepted a job at a men's sexual dysfunction clinic. 8:30 to 5, Monday to Friday, no weekends. Carte blanche for research. The only RN on staff.

It was time for some reduced stress. It was time to close down the hotel sewing machines and work professionally again after four years in the wilderness.

END NOTES

The adventures at my first RN job are written up as *The Weenie Clinic*. Visit my web site (www.henryintili.com) for more information.

Autographed copies of this book and my other books are available at www.henryintili.com.

Henry Intili

Write the author:
info@henryintili.com

www.ingramcontent.com/pod-product-compliance
Lightning Source LLC
Chambersburg PA
CBHW030805180526
45163CB00003B/1155